Wood Fracture Characterisation

Wood Fracture Characterisation

Marcelo F. S. F. de Moura and Nuno Dourado

CRC Press is an imprint of the
Taylor & Francis Group, an **informa** business

MATLAB® is a trademark of The MathWorks, Inc. and is used with permission. The MathWorks does not warrant the accuracy of the text or exercises in this book. This book's use or discussion of MATLAB® software or related products does not constitute endorsement or sponsorship by The MathWorks of a particular pedagogical approach or particular use of the MATLAB® software.

CRC Press
Taylor & Francis Group
6000 Broken Sound Parkway NW, Suite 300
Boca Raton, FL 33487-2742

© 2018 by Taylor & Francis Group, LLC
CRC Press is an imprint of Taylor & Francis Group, an Informa business

No claim to original U.S. Government works

Printed on acid-free paper

International Standard Book Number-13: 978-0-8153-6471-9 (Hardback)

This book contains information obtained from authentic and highly regarded sources. Reasonable efforts have been made to publish reliable data and information, but the author and publisher cannot assume responsibility for the validity of all materials or the consequences of their use. The authors and publishers have attempted to trace the copyright holders of all material reproduced in this publication and apologize to copyright holders if permission to publish in this form has not been obtained. If any copyright material has not been acknowledged please write and let us know so we may rectify in any future reprint.

Except as permitted under U.S. Copyright Law, no part of this book may be reprinted, reproduced, transmitted, or utilized in any form by any electronic, mechanical, or other means, now known or hereafter invented, including photocopying, microfilming, and recording, or in any information storage or retrieval system, without written permission from the publishers.

For permission to photocopy or use material electronically from this work, please access www.copyright.com (http://www.copyright.com/) or contact the Copyright Clearance Center, Inc. (CCC), 222 Rosewood Drive, Danvers, MA 01923, 978-750-8400. CCC is a not-for-profit organization that provides licenses and registration for a variety of users. For organizations that have been granted a photocopy license by the CCC, a separate system of payment has been arranged.

Trademark Notice: Product or corporate names may be trademarks or registered trademarks, and are used only for identification and explanation without intent to infringe.

Visit the Taylor & Francis Web site at
http://www.taylorandfrancis.com

and the CRC Press Web site at
http://www.crcpress.com

Contents

Preface vii

Acknowledgements ix

Authors xi

1 Introduction 1
   1.1   Global Overview and Applications 1
   1.2   Wood – How Does It Form? 5
   1.3   Wood Constituents and Micro-Structure 7
   1.4   Wood at the Mesoscale 10
   References 14

2 Wood Mechanical Behaviour 15
   2.1   Elastic and Strength Properties 15
         2.1.1   Young's Moduli and Normal Strengths 16
         2.1.2   Poisson's Ratios 20
         2.1.3   Shear Moduli and Strengths 20
   2.2   Strength Failure Criteria 26
   2.3   Fracture Mechanics Based Approaches 29
         2.3.1   Linear Elastic Fracture Mechanics 29
         2.3.2   Cohesive Zone Models 32
   References 35

3 Mode I Fracture Characterisation 37
   3.1   Double Cantilever Beam 37
         3.1.1   Test Description 37
         3.1.2   Classical Data Reduction Schemes 37
         3.1.3   Modified Experimental Compliance Method (MECM) 39
         3.1.4   Compliance-Based Beam Method (CBBM) 40
         3.1.5   Numerical Validation 42
         3.1.6   Experimental and Numerical Results 45
   3.2   Single-Edge-Notched Beam Loaded in Three-Point-Bending 47
         3.2.1   Test Description 47
         3.2.2   Data Reduction Scheme Based on Equivalent LEFM 49
         3.2.3   Compliance-Based Beam Method 51
         3.2.4   Numerical Validation of the Compliance-Based Beam Method 53
         3.2.5   Experimental and Numerical Results 55
   3.3   Tapered Double Cantilever Beam 55
         3.3.1   Test Description 55
         3.3.2   Data Reduction Scheme 55
         3.3.3   Compliance-Based Beam Method 58
         3.3.4   Numerical Validation 60
         3.3.5   Experimental and Numerical Results 61
   3.4   Compact Tension Test 62
   3.5   Conclusions of Mode I Fracture Tests 64
   References 64

v

vi   *Contents*

4 Mode II Fracture Characterisation 67
   4.1   End-Notched Flexure Test 68
       4.1.1   Test Description.. 68
       4.1.2   Classical Data Reduction Schemes.................................. 69
       4.1.3   Compliance-Based Beam Method 70
       4.1.4   Experimental and Numerical Results............................... 72
   4.2   End-Loaded Split Test (ELS) 74
       4.2.1   Test Description.. 75
       4.2.2   Classical Data Reduction Schemes.................................. 76
       4.2.3   Compliance-Based Beam Method 77
       4.2.4   Experimental and Numerical Results............................... 78
   4.3   Four End-Notched Flexure Test 79
       4.3.1   Test Description.. 81
       4.3.2   Compliance Calibration Method 81
       4.3.3   Compliance-Based Beam Method 82
       4.3.4   Experimental and Numerical Results............................... 84
   4.4   Conclusions of Mode II Fracture Tests...................................... 87
   References 87

5 Mixed-Mode I + II Fracture Characterisation 89
   5.1   Single-Leg Bending Test 90
       5.1.1   Test Description.. 90
       5.1.2   Compliance-Based Beam Method 91
       5.1.3   Numerical Analysis... 91
       5.1.4   Experimental Results 93
   5.2   End Load Shear-Mixed Mode Test 93
       5.2.1   Test Description.. 93
       5.2.2   Compliance-Based Beam Method 94
       5.2.3   Numerical Analysis... 95
       5.2.4   Experimental Results 96
   5.3   Mixed-Mode Bending Test 97
       5.3.1   Test Description.. 97
       5.3.2   Experimental Analysis and Results.................................. 99
       5.3.3   Numerical Validation 102
   5.4   Conclusions of Mixed-Mode I + II Fracture Tests 103
   References 103

6 Structural Applications – Case Studies 105
   6.1   Wood Bonded Joints 105
       6.1.1   Repaired Beam under Tensile Loading 105
       6.1.2   Repaired Beam under Bending Loading.......................... 109
       6.1.3   Reinforcement of Wood Structures.................................. 113
   6.2   Wood Dowel Joints 117
       6.2.1   Steel–Wood–Steel Connection 117
       6.2.2   Wood–Wood Joint.. 122
   6.3   Conclusions of Structural Applications.................................... 129
   References 130

Index 133

Preface

Wood is increasingly used in structural applications. In fact, the interest in applying renewable resources in structural design is growing due to ecological and environmental reasons, and energy shortages. Solid wood is generally used in frames, buildings, truss roof structures in buildings, bridges, towers, railroad infrastructures and in many other applications. Damage and failure behaviour of wood members in tensile, compressive or shear loading are extremely important to account in wooden structures subjected to high working stresses. Structural details involving wood member's connections also require special attention for a safe design. Effectively, damage under tensile, compressive or shear loading can occur at the joints or within the lumber of many members. The criteria currently used in wood structural design are based on stress or strain analyses. However, many structural applications of wood involve discontinuities and singularities such as notches or holes, which lead to important stress concentration effects. Additionally, wood as a natural and biological material presents drastic variations in its inner structure as a result of internal defects such as knots, variation of grain orientation, reaction wood and others that contribute to a considerable source of variability at several levels. The consequence of these aspects is the consideration in the actual design codes of several and high safety factors in the design of wood structures.

To overcome these drawbacks and limitations, a promising line of research consists in employing fracture mechanics concepts to wood design. Such methodology can contribute significantly to a better understanding and more reliable design methods concerning the project of wood structural applications. Therefore, a comprehensive description of the fundaments on fracture mechanics is presented in this book, as well as the necessary extensions that account for wood specificities. In this context, a wide description and development of several fracture tests appropriate for wood fracture characterisation under different loading modes is the main focus of this book. The described work includes new fracture tests applied to wood, new data reduction schemes and numerical models based on cohesive zone analysis that are frequently used to validate the proposed experimental-based methodologies. In addition, several structural details involving connections of wood members have been analysed for safe design in the last chapter, using the experimental and numerical tools described throughout the book.

In summary, this book addresses the following main targets:

- Proposal of accurate methods for the design of wood members in structures;
- Development of approaches that allow decreasing the material consumption in structural applications;
- Enhancement of the attractiveness of wood for structural application purposes;
- Stimulation of the building sector to employ wood as a competitive construction material, that allows boosting the forestry and woodworking industry sectors;
- Valuable contribution towards ecological and environmental issues.

The main goal of this book is to stimulate the readers to use the theories and methods described herein as tools for safe, ecological and efficient wood structural design.

vii

viii *Preface*

MATLAB® is a registered trademark of The MathWorks, Inc. For product information, please contact:

The MathWorks, Inc.
3 Apple Hill Drive
Natick, MA 01760-2098 USA
Tel: 508-647-7000
Fax: 508-647-7001
E-mail: info@mathworks.com
Web: www.mathworks.com

Acknowledgements

The authors thank the Portuguese Foundation for Science and Technology for supporting the work here presented through the projects 'Design of wood-bonded joints' (POCTI/EME/45573/2002), 'Design of wood and wood-bonded joints under mixed-mode loading' (POCI/EME/56567/2004) and 'Repair of wood structures using artificial composites' (PTDC/EME-PME/64839/2006).

The authors also thank Professor José Joaquim Lopes Morais from the University of Trás-os-Montes e Alto Douro for his advices, ideas and participation in the generality of the topics. All the students participating in the projects are also recognised. Professors Stéphane Morel and Gérard Valentin from the University of Bordeaux are also acknowledged. The authors would like to highlight and express their gratitude to the students Manuel António Lima da Silva, Jorge Marcelo Quintas de Oliveira and João Pedro da Costa Reis for their good spirit and valuable experimental and numerical work that contributed decisively to the presented work. The authors are also grateful to Professor José Lousada for his courtesy in providing the wood histological images.

Authors

Marcelo F. S. F. de Moura is an associate professor with aggregation at the Department of Mechanical Engineering of the Faculty of Engineering of the University of Porto, Portugal. His research interests are focused on mechanical and fracture behaviour of anisotropic materials (composites, wood and bone) and adhesive bonding. Numerical simulation of fracture and fatigue using cohesive zone modelling is a prominent research topic. He is author/co-author of 150 research papers in international scientific journals and 2 books, and he has participated in 26 research projects, being leader of 8 of them and supervised 8 PhD (2 ongoing).

Nuno Dourado is an assistant professor at the Mechanical Department of the School of Engineering of the University of Minho in Guimarães, Portugal. His research interests are the fracture characterisation of quasi-brittle materials (wood, concrete and bone tissue) and cohesive zone modelling of these materials. The characterisation of viscoelastic response of biological materials is a recent topic of research. He has authored/co-authored 45 research papers, participated in 7 scientific projects (1 ongoing), and supervised/co-supervised 26 MSc Thesis and 6 PhD (5 ongoing).

1

Introduction

1.1 Global Overview and Applications

Wood is a natural renewable material that has been used by humanity for numerous daily functions since ancestral times. Its capacity to be cut into different shapes (Figure 1.1a), its hard-wearing aptitude, superior mechanical resistance, abundance in many latitudes, reasonable price and attractiveness are the features that have contributed to its constant search. Notable machinability, lightweight and durability in many environmental conditions are relevant characteristics that have contributed to its use, as well. The fact that wood is prone to change its form under specific environmental conditions (i.e. moisture, temperatures and loading) has also enabled the development of many artefacts of refined geometrical configuration. These characteristics are identified through many applications such as domestic furniture or objects (Figure 1.1b), sport products (Figure 1.1c) and transport vehicles (e.g. ships, trains, motor vehicles and aircrafts) (Figure 1.1d). Wood is also recognised as a naturally beautiful and aesthetically pleasing material.

Other arguments have recently been raised, stirring up the employment of wood (and timber) in structural applications, as is the case of bridges (Figure 1.2a), monuments (Figure 1.2b) and ceilings (Figure 1.2c).

In particular, the application of wood in the building construction sector is a way to increase carbon sequester capacity and reduce energy consumption. In fact, wood is a naturally grown and engineered material that contributes to an overall elimination of greenhouse gases from the atmosphere. This occurs because wood products store the carbon that growing trees have sequestered from the air during its lifetime (photosynthesis). In addition, due to the referred good machinability, production and processing of wood products require much less energy than that necessary for most other traditional building materials. This important feature gives wood products a significantly lower carbon footprint, which results from the substitution of materials that require larger amounts of fossil fuels to be produced. Another aspect is related to the fact that wood can offer quicker build times, with more innovative design approaches than traditional materials, benefiting from the fact that woodwork is often done on site or pre-fabricated locally or regionally. This aspect is also pointed out as a way to support local jobs for carpenters and craftspeople. Other aspects that can be mentioned refer to both excellent thermal and acoustic insulation properties of wood. These characteristics are highly valued by architects and civil engineers as they concern the achievement of increasingly demanding prerequisites for buildings certification. As a result, in many developed countries, wood is taking over from concrete and steel as the architectural marvel material of the twenty-first century. Concerns about environmental changes have already led some of those countries to embrace the construction of wooden high-rise buildings (Figure 1.3).

1

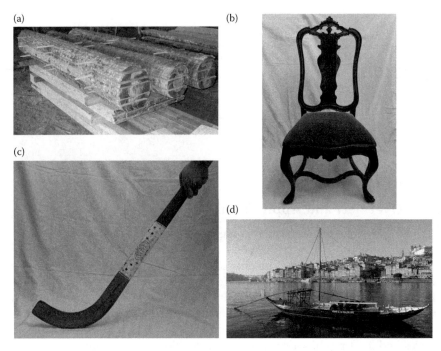

FIGURE 1.1
(a) Timber frames, (b) wooden furniture (chair), (c) wooden hockey stick, (d) Port wine carrying boats (Rabelo).

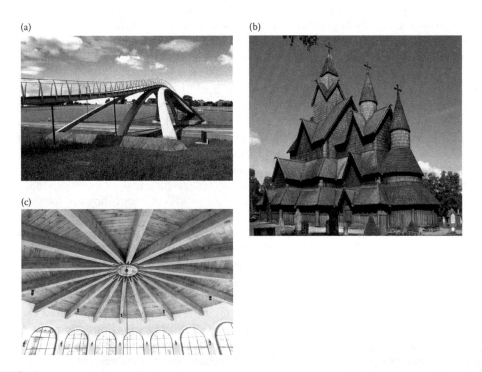

FIGURE 1.2
(a) Leonardo bridge in Aas, Norway (courtesy of Robert Cortright/Bridge Ink), (b) church (Heddal stavekirck, Norway), (c) ceiling. (Photography of Blake Bronstad.)

Introduction

FIGURE 1.3
Puukuokka building, a wooden eight-storey apartment block in Finland. (Courtesy of Mikko Auerniitty.)

Wood also represents an important source of energy (transformed in pallets or unchanged), which is used for heating and for the production of electricity. Many countries use important quantities of wood waste and sawdust that have been gathered in forests and wood processing industries to produce energy in biomass technological facilities. This is also the destination for many wooden structures at the end of their life cycle. It is stated that the environmental impact ensued by carbon emissions during the combustion balances the quantity of carbon that has been sequestered in the life cycle of the tree. Nowadays, wood remains the raw material for a wide variety of current products, although other competitive materials are available. For many traditional uses, the importance of wood remains unchanged, rising gradually in new products to meet actual needs of humanity.

Subsequent to harvesting in the forest, wood can be transformed into many different products. Numerous industrial processes like sawing, board cutting, gluing, fragmentation, pulping, pigmentation or processing with different chemical agents accomplish this. Among the industrial wood products are plywood (Figure 1.4a), medium-density fibreboard (Figure 1.4b), oriented strand board (Figure 1.4c), glued laminated timber (glulam) (Figure 1.4d), laminated veneer lumber (Figure 1.4e), cross-laminated timber (Figure 1.4f), parallel strand lumber (Figure 1.4g), laminated strand lumber (Figure 1.4h), particleboard (Figure 1.4i), I-joists and wood I-beams (Figure 1.4j), finger-jointed lumber (Figure 1.4k), paper and many others. These products are strong examples of the wood processing ability that contribute decisively to its wide application in several fields of industry.

FIGURE 1.4
(a) Plywood, (b) medium-density fibreboard (MDF), (c) oriented strand board (OSB), (d) glued laminated timber (glulam), (e) laminated veneer lumber (LVL), (f) cross-laminated timber (CLT), (g) parallel strand lumber (PSL), (h) laminated strand lumber (LSL), (i) particleboard, (j) wood I-beam and (k) finger-jointed lumber.

Introduction 5

1.2 Wood – How Does It Form?

Wood is a product ensuing from the trees. Trees as vascular plants are organised into three major organ parts, according to its key functions: root, stem and leaves (Figure 1.5). Roots provide a complex anchorage system that enables fixing the tree to the ground, while assuring the absorption of water and mineral resources from the soil (i.e. constituents of the sap). In some instances, roots are able to enhance the nutrition capacity of the tree by accomplishing symbiotic associations, using nitrogen-fixing microorganisms and fungal symbionts known as mycorrhizae. This reciprocal relation enables increasing phosphorous uptake to the tree. In seasonal climates, roots also serve as nutrient storage depots.

The stem, containing (normally) the majority of the substances of a tree, is responsible for conducting the sap upwards from the roots to the leaves, providing the structural support to the leaves (crown) and storing nutrients. This transportation of the sap is assured by the tracheids (i.e. elongated cells or fibre cells that serve in the transport of water and mineral salts) through the sapwood, sited between the cambium (single layer of actively

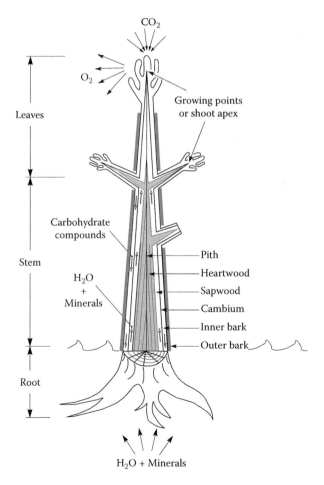

FIGURE 1.5
Illustrative representation of a tree showing the conoid layers along the stem. (Adapted from Smith et al., 2003.)

dividing cells responsible for wood secondary growth) and heartwood (Figure 1.5). Sap transportation also occurs perpendicular to the pith, a primary tissue in the form of a central parenchymatous cylinder. Sap runoff in the transverse plane of the tree assures the radial transportation of nutrients from the sapwood to the heartwood. Tree growth results from the continuous superposition of conoid layers up to the growing points (Figure 1.5) produced by the cambium (Figure 1.6), which in many species turns the older tissues less important to assure physiological processes (i.e. transportation of nutrients). These tissues are gradually converted by the tree in heartwood, which has no intrinsic active defences against infections (died cells). For these reasons, xylem vessels (sapwood and heartwood; Figure 1.5) are blocked with tyloses (i.e. natural structures that dam up the sap), gums, resins and waxes. These substances have high concentrations of volatile organic compounds as terpenes that are toxic to insect larvae and tree pathogens, such as bacteria and fungi, which in several trees turn heartwood darker than sapwood. Examples of dark heartwood are the pine (e.g. *Pinus pinaster*), douglas fir (*Pseudotsuga menziesii*), redwood (*Sequoia sempervirens*), eastern hemlock (*Tsuga canadensis*), cypress (*Cupressus*), oak (*Quercus*), walnut (*Juglans regia*), chestnut (e.g. *Castanea sativa*), elm (*Ulmus minor*) and others (Tsoumis, 1991). This difference in the colour of heartwood relative to sapwood is not visible in all wood species, although heartwood is present in all trees once a certain age is attained. As the diameter of the stem increases, older (annual) rings progressively stop participating in the life processes of the tree, providing only mechanical support. The tissue that surrounds the stem is the bark (Figure 1.6), which differs in texture and thickness according to the species and age. Growth layers piled in the bark are not macroscopically defined. In older trees, two tissues are identified: inner bark (light colour, thin and humid) and outer bark (dark, dry and corky). The outer layers of the inner bark progressively change into outer bark, following a process similar to the renovation that occurs during the formation of the heartwood. The outer layers of the outer bark progressively fall off.

Leaves in many trees consist of a broad, extended blade (the lamina), that are attached to the plant twig or stem by the petiole. Generally, leaves are widely vascularised organs formed by networks of vascular bundles that contain xylem, which assure water supply for photosynthesis, and phloem, a substance containing sugars formed by the photosynthesis. Leaves are fairly diverse in size and shape, which include the arrangement of the veins that supports the lamina (broad expanded part) and provides the transportation of

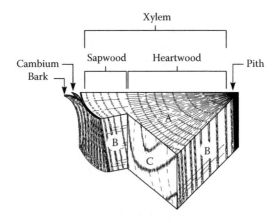

FIGURE 1.6
Schematic representation of a wood section displaying (A) the transverse, (B) radial, and (C) tangential surfaces. (Adapted from Tsoumis, 1991.)

Introduction 7

materials to and from other tissues. Leaves may be simple (unique blade) or compound (several blades); a few assume the form of a spine or scale. The main function of the leaves is to produce nutrients for the plant by photosynthesis. This is achieved by light energy absorption through the chlorophyll, the photosynthetic pigment that gives plants their typical colour (mostly green or greenish). Approximately one-fifth of the central leaf is constituted by specialised subunits that absorb sunlight, the chloroplast, which together with certain enzymes enable the decomposition of the water molecule into its basic elements, hydrogen and oxygen. The energy conversion and corresponding storage are accomplished in energy-storage molecules adenosine triphosphate (ATP) and nicotinamide adenine dinucleotide phosphate, respectively. Chloroplasts use these molecules to make organic molecules (sugars) from carbon dioxide captured from the atmosphere, in a process known as the Calvin cycle. These organic molecules are basic nutrients of the plant, which are essential to herbivores. Through this biochemical process, the oxygen liberated from green leaves (from pores called stomates) restores the oxygen that is consumed from the atmosphere by living beings and combustion (Figure 1.5). This phenomenon is one of the important aspects that makes wood an appealing material for industrial applications.

1.3 Wood Constituents and Micro-Structure

Wood can be viewed as a natural composite material and a biochemical complex of cellulose, hemicelluloses, lignin and extractives (Tsoumis, 1991). Cellulose is a complex carbohydrate (or polysaccharide) formed by glucose units, which constitutes the basic structural component of cellular walls. This constituent presents a crystalline microstructure, offering mechanical strength to cell walls and resistance to hydrolysis. This compound comprises 40%–50% of wood in the form of microfibrils (i.e. bundles of cellulose). Hemicelluloses are the matrix constituents that surround (with other carbohydrates) the cellulose fibres (microfibrils) of plant cells (Hon and Shiraishi, 2001). Hemicelluloses are low-molecular-weight short-length mixed polymers that contribute to strengthen the cellular walls. Bounded to the surface of cellulose network, this heterogeneous polymer prevents direct contact of microfibrils, forming a gel phase. Hemicelluloses do not originate aggregates; however, they play an important role in the mechanical properties of wood by bonding with both lignin and cellulose. Molecules of hemicellulose are particularly important in the regulation of the rate at which primary cell walls expand during its growth and do not have chemical association to cellulose. Lignin is the encrusting substance solidifying the cell wall associated with matrix substances. It is a significant part of all vascular plants, taking part in the formation of the cell wall subsequent to cellulose (40%–50%), hemicellulose (15%–25%) and polysaccharides (Lawoko et al., 2006; Gellerstedt, 2015). Lignin is a complex three-dimensional organic polymer that limits the penetration of water into wood cells and makes wood compact, conferring mechanical strength and stiffness to the cellular wall. In dry weight state, lignin makes up 15–40 wt% of woody plants (Baurhoo et al., 2008; Dohertya et al., 2011).

Wood extractives are chemical compounds that may be extracted from wood, bark or foliage using various neutral solvents (e.g. dichloromethane, methanol, petroleum ether, acetone, ethanol, water). Those substances may be lipid extractives (aliphatics, terpenoids), phenolic extractives (simple phenolics, stilbenes, flavonoids, lignans) and other extractive compounds (alkanes, proteins, monosaccharides and derivatives). The quantity of extractives that wood can contain varies with the wood species and the position within the tree, being in the range of 1%–20% (Walker et al., 1993). Wood extractives act as antifeedants, antioxidants, antivirals, bacteriacides and fungicides, with their presence accounting for the colour of heartwood (Figure 1.6) (black in ebony; red-brown in cedars, douglas fir and hemlock; yellow orange in southern pine; or indistinguishable in pale sapwood with poplar).

Water is also one of the important constituents of wood. Water in wood may exist in the form of free water and/or bound water. Free water can be found in cell lumens (Figure 1.7) and inside cell hollows, in the state of vapour and/or liquid. Bound water is part of the cell wall material, which is in the form of hydrogen bonds and van der Waals forces (Dumitriu, 2004), being more difficult to remove than free water. When free water (or absorbed water) moves out by liquid flow or vapour diffusion, leaving the wood substance completely saturated with bound water, wood cells reach what is commonly called the fibre saturation point (FSP). Though in this state wood remains fully saturated, no water exists in the cell

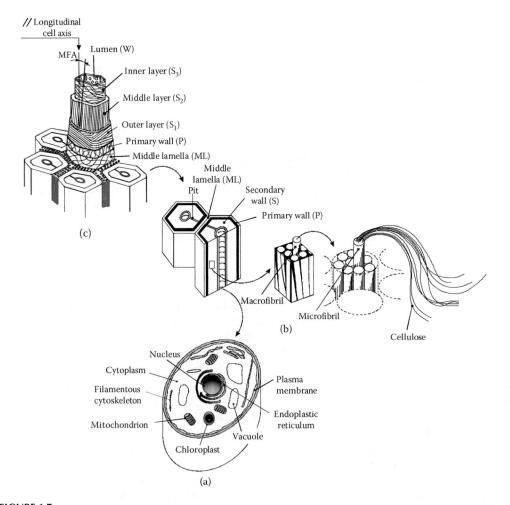

FIGURE 1.7
(a) Schematic of a plant cell, (b) the fibrillar structure of cell wall and (c) structure of a cellulose fibre unit showing the mature cell wall layers: middle lamella (ML), primary wall (P), secondary wall (S) [outer layer (S_1), middle layer (S_2) and inner layer (S_3)] and lumen (W).

Introduction 9

lumen (Figure 1.7), meaning that no more water can be held between the microfibrils. Hence, excluding the water quantity that is present in wood constituents (i.e. bound water), which is only removed by means of destructive methods, water in wood is retained in cell walls. When wood is harvested and the seasoning process is initiated, a significant loss of bound water (drying process) occurs in a rate that is governed by the relative humidity (RH) of the involving atmosphere.

The plant cell (Figure 1.7a) constitutes the elementary unit of the wood structure, being the smallest living entity (10–100 μm) able to function in an independent way. The cell has many roles, as manufacturing proteins, polysaccharides and providing the deposit for minerals. Unlike animal cells, the cell walls in plants are placed outside the plasma membrane (with 0.1–100 μm thick), which turns plant cells more rigid and capable of withstanding internal pressure as high as 1 MPa. The concentration of plant nutrients and retention of synthesised products within the cell (to assure its vital functions) are enabled by the plasma membrane (Figure 1.7a). This biological structure is composed of amphipathic molecules, i.e. exhibit hydrophilic (water liking) and hydrophobic (water disliking) behaviour, which renders it possible to function as a selective barrier. The most important differentiated structure within a cell is the nucleus. It contains the genetic information (i.e. the DNA) that is essential to undergo the control of the cell structure and function. The involving region is filled with fluid (cytosol) that synthesises proteins within a substance known as endoplasmic reticulum and Golgi apparatus (i.e. glycosylation enzymes that attach various sugar monomers to proteins). Other cellular organelles (i.e. specialised biological subunits) found within the plant cells are the chloroplasts and mitochondria, which play an important role in the energy-converting systems from sunlight and energy storage, respectively. The remaining organelles are the vacuoles, huge vesicles that are responsible for the storage of metabolites (compounds synthesised by plants), nutrients and even waste products ensued by the plant metabolism.

In an early stage of the woody cell differentiation (Figure 1.7b), the living protoplasm produces the primary wall (P) that can be widely enlarged in its surface as the cell wall develops. The substance that exists in the interface of the primary walls of adjacent woody cells is designated as intercellular layer (I) or middle lamella (0.2–1.0 μm thick). Since it is difficult to differentiate the region between the I layer and the P wall in the mature cell, the term compound middle lamella is generally used when referring to the combined I layer and the two adjacent P walls. With the completion of wall formation, the cell wall layers are made by the apposition of wall substances onto the inside of the primary wall (P) (Figure 1.7c). These wall layers are designated by secondary wall (S) (10–20 μm thick), being formed by three layered structures with cellulose microfibrils differently oriented (Figure 1.7c), predominantly in a parallel manner forming fibrillary bundles (fibril aggregates). These layers are composed of a relatively narrow or thin outer layer (S_1), an inner layer (S_3) and a relatively thick middle layer (S_2). The three layers of the so-called secondary wall (S), known as S_1, S_2 and S_3, are disposed in a plywood type of construction. The middle layer (S_2) of the secondary wall is the thickest. Therefore, S_2 is the one that contributes most to the bulk of the cell wall material, forming a compact region characterised by the existence of a high degree of parallelism of microfibrils. For this reason, it is reported that S_2 layer has a greater influence on physical and mechanical properties of wood (Cave, 1997). The S_3 layer is a thin layer of flat helices presenting an orientation not very different of S_1, though loosely textured and poorly developed. Although S_2 exhibits a microfibrillar orientation with steep helices (Figure 1.7c), there are transition lamellae on its outer and inner surfaces. Numerous lamellae in these regions present a gradual shift of microfibril angles (MFAs) between S_1 and S_2 and between S_2 and S_3. However, the gradual shift of MFA is more abrupt between S_2 and S_3 than between S_1 and S_2. MFA refers to the orientation of the cellulose bundles relative to the longitudinal cell axis (Figure 1.7c). The thin primary wall (P) consists of a loose aggregation of microfibrils oriented more or less axially to the longitudinal axis of the cell on the outer surface. Cell walls are not totally complete around its perimeter. They are interrupted by narrow pits (Figure 1.7b) that allow the exchange of fluids between adjacent cells. As independent living entities, plant cells exhibit a high level of self-organisation and assembly.

From the material microscale viewpoint, wood is essentially a two-phase material constituted by layers in cell walls. These layers are formed by strong and flexible crystalline cellulose fibres (Figure 1.7b) surrounded and held together by a stiffer matrix containing non-crystalline cellulose, hemicellulose and lignin. All the constituents of this remarkable organism undertake continuous renewal and adaptation to environmental conditions, making the tree a dynamic system.

1.4 Wood at the Mesoscale

The most common shape of the transverse section of a tree stem is circular, with the referred three parts being identified as pith, xylem and bark (Figure 1.6). Cambium can be observed by means of a microscope between xylem and bark, being responsible for production of both. Pith is normally at the centre of the stem. In certain species, the pith is very small, being hardly visible with the naked eye. Other species, such as red elder (*Sambucus microbotrys*) and tree-of-heaven (*Ailanthus altissima*) exhibit visible large piths. Pith is reasonably uniform in softwood, contrasting with hardwood,* which reveals different shape, colour and structure. Pith is star shaped in oak; triangular in beech (*Fagus sylvatica*), birch (*Betula papyrifera*) and alder (*Alnus rubra*); ellipsoid in basswood (*Tilia americana*), ash (*Fraxinus americana*) and maple (*Acer campestre*); circular in walnut (*Juglans nigra*), elm (*Ulmus americana*) and willow (*Salix babylonica*); squarish in teak (*Tectona grandis*) (Jane, 1970). Pith colour may vary from black to whitish, with a structure that may be continuous (solid), spongy (porous), chambered or hollow (Harlow and Harrar, 1991).

Concentric layers, or growth rings, are visible in the transverse section of the stem (Figure 1.6). They present a characteristic pattern, which results from the referred mechanism of the tree growth, based on the superposition of structurally different conoid layers (Figure 1.5). In temperate zones, a wood layer (a bark layer as well) is added every growth season. In stark contrast with this observation, growth rings are not always distinct in tropical zones. When visible, they usually tally with alternating wet and dry periods (i.e. rainfall seasonability). However, for the majority of wood species, growth rings are easily noticed from each other due to obvious differences between earlywood and latewood (Figure 1.8). These wood tissues (respectively known as springwood and summerwood) vary considerably from each other regarding their cellular structure (cell wall thickness and cell size), which reflects on wood local density and colour. These two tissues are not

---

* Softwoods are formed by conifers (gymnosperms, e.g. hard pines, douglas fir, Mediterranean cypress and yew), while hardwoods are produced by broad-leaved species (angiosperms, e.g. willow, poplar, basswood and balsa). Softwoods are known to retain its leaves throughout the year and expose their seeds, while hardwoods shed their leaves annually and provide a cover for their seeds. The primary distinguishing feature between softwoods and hardwoods is that softwood species lack vessels or pores.

Introduction

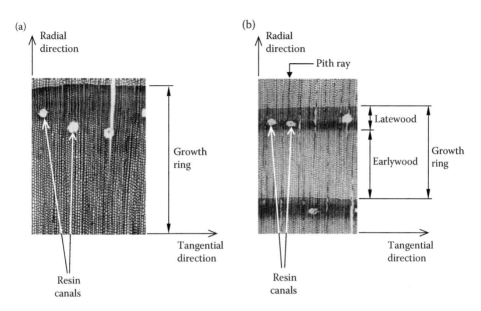

FIGURE 1.8
Different transitions in softwood between earlywood and latewood: (a) gradual in picea (*Picea abies* L.) and (b) abrupt in maritime pine (*Pinus pinaster* Ait.).

readily identifiable if the transition is gradual (Figure 1.8a), or are clearly visible in those cases where the transition of the cell size is abrupt (Figure 1.8b). Due to seasonal changes in growth, wood fibres (longitudinal tracheids aligned with the pith) formed during the rapid cell development in spring (earlywood fibres) present thin walls, while fibres produced later in the growth period in summer (latewood fibres) are thick walled. Thus, earlywood and latewood fibres formed within a year constitute the annual ring or growth ring. In softwoods, latewood is darker and denser than earlywood. Contrasting with this feature, less pronounced differences in their microscopic structure are noticed in hardwoods. The most relevant characteristic of softwoods is the presence of pores, which correspond to small roundish canals that penetrate the growth ring. Intercellular canals are spaces in wood tissue. These anatomic structures are not cells, but rather lined arrangements with specialised parenchyma cells called epithelial cells. Intercellular canals may occur in both softwoods (resin canals) and hardwoods (gum canals), being rarely visible in species of the temperate zone (Tsoumis, 1991). These ducts may be visible with the naked eye or by means of a magnifying lens. This anatomic feature enables classifying hardwoods into two large groups: (a) ring-porous, with earlywood exhibiting large porous disposed around the pith (e.g. oak and chestnut, visible in Figure 1.9a) and (b) diffuse-porous, which according to the name prompts to fairly uniform and scatter porous (e.g. beech or poplar shown in Figure 1.9b). Another obvious feature of softwood lies in its radial alignment in the cross-section plane which makes softwood species much more anisotropic in this plane than hardwood.

Clearly, growth rings are easier to distinguish in ring-porous wood arrangements than in diffuse-porous woods (Figure 1.9a,b).

The earlier described macroscopic characteristics of wood presume a normal tree growth. Hence, the presence of numerous defects in the tree can significantly modify the appearance, in particular the radial and tangential surfaces, and change the material

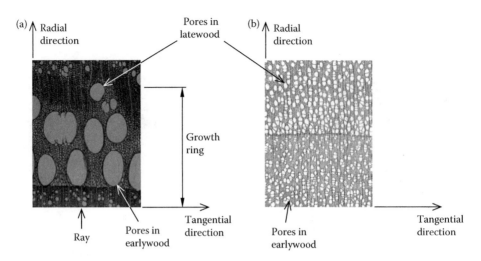

FIGURE 1.9
Different arrangement of pores in hardwood: (a) ring-porous in chestnut tree (*Castanea sativa* Mill.) and (b) diffuse-porous in poplar (*Populus* spp.).

mechanical behaviour. Another aspect that alters the mechanical behaviour of wood is the water quantity. In fact, high moisture gradients in wood frequently lead to damage initiation and permanent deformations in wooden members. Consequently, extreme care should be taken to control the atmospheric conditions in the course of the drying process. In the UK and most regions of USA, the air-drying timber seasoned in most adequate conditions is accomplished at 15%–18% moisture (Umney and Rivers, 2003). The condition that results from the establishment of a balance between the amount of bound water present in wood with the RH in the surrounding atmosphere is called the equilibrium moisture content (EMC). The typical trend of the EMC as a function of RH for wood is schematically shown in Figure 1.10 (for 21°C). The plotted dashed region represents the envelope where the curves of different wood species will be depicted. The thick line represents the average data for white spruce with a FSP around 28%. The FSP differs somewhat among the wood species. Hence, for woods that present high extractive quantities (e.g. western larch, rosewood or mahogany), the FSP can be as low as 22%–24%. Others, like those with low extractive quantities (e.g. beech, birch or *eucalyptus grandis*), the FSP reaches the range 32%–35%. The EMC is affected by the temperature. Hence, at intermediate humidity levels the EMC will be reduced by 1% for every increase of 1°C–4°C (Umney and Rivers, 2003).

The amount of moisture present in wood is usually quantified as a percentage of the weight of the wood when oven dry. The most common methods are the oven-drying (OD) and electrical methods. The former is the most universally accepted, though it is time consuming and requires cutting of wood samples. However, OD method may slightly overestimate moisture content (MC) if wood contains volatile extractives that remain in the material. On the other hand, the electrical method provides rapid measurements, is non-destructive and therefore is particularly applicable to in-service timber structures. However, the interpretation of the ensued results has to be performed with caution and the values of MC cannot exceed 30%. For this reason, the OD method is the one that is more commonly reported in the scientific and technical literature.

Introduction

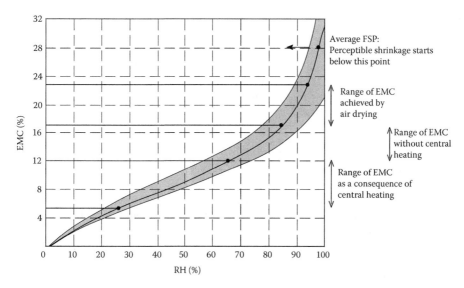

FIGURE 1.10
Average relationship between the equilibrium moisture content (EMC) of wood and relative humidity (RH) of the surrounding atmosphere at 21°C. Fibre saturation point (FSP; average 28% of EMC) and dry rot safety line. (Adapted from Umney, 2003.)

MC of a given specimen may be determined on a cross section taken from the test piece. For the characterisation of wood for timber structural applications (e.g. beams), the analysed sample may be of full cross section, free from knots and resin pockets. Hence, the wood section has to be weighed in advance (mass, m_1) and then oven dried at a temperature of 103°C ± 2°C, till no appreciable weight changes is noticed in 4 hr waiting intervals. In these conditions, the recorded wood mass (m_2) is considered constant and MC has converged to zero. The waiting time to record the final mass (m_2) will depend on the volume of the sample. Then, MC will be evaluated as follows:

$$\text{MC}(\%) = \frac{m_1 - m_2}{m_2} \times 100 \tag{1.1}$$

The FSP stated has an MC value, being characteristic of a given wood species, usually in the range of 25%–30%. Frequently, sapwood presents higher values of MC than heartwood in softwoods. This is not perceived in hardwoods since it varies from one species to another.

The physical properties of wood strongly depend on the quantity of water that is present in the material. Hence, when establishing procedures to measure a given physical property, it is vital to state the environmental conditions that have to be maintained (i.e. temperature and relative humidity) to perform the corresponding measurements. Furthermore, the observation of the environmental conditions where a timber structure is to be built is of crucial importance, since wood will undergo dimensional changes due to intrinsic hygroscopic dynamic mechanisms that will affect its structural performance. In the subsequent chapters of this book, ambient conditions, i.e. temperatures ranging between 20°C and 25°C and relative humidity between 60% and 65%, have been considered during the experimental tests.

References

Baurhoo, B., C. A. R. Ruiz-Feria and X. Zhao (2008). Purified lignin: Nutritional and health impacts on farm animals-A review. *Animal Feed Sci Technol*, 144:175–84.

Cave, I. D. (1997). Theory of X-ray measurement of microfibril angle in wood. Part 1. The condition for reflection. X-ray diffraction by materials with fibre type symmetry. *Wood Sci Technol*, 31:143–52.

Dohertya, W. O. S., P. Mousaviouna and C. M. Fellows (2011). Value-adding to cellulosic ethanol: Lignin polymers. *Ind Crop Prod*, 33:259–76.

Dumitriu, S. (2004). *Polysaccharides, Structural Diversity and Functional Versatility*, Second Edition, CRC Press, New York. ISBN 13: 978-1-4200-3082-2.

Gellerstedt, G. (2015). Softwood kraft lignin: Raw material for the future. *Ind Crop Prod*, 77:845–54.

Harlow, W. M. and E. S. Harrar (1991). *Textbook of Dendrology (American Forestry)*, McGraw-Hill Book Co., New York. ISBN 10: 0070265712; ISBN 13: 9780070265714.

Hon, D. N.-S. and N. Shiraishi (2001). *Wood and Cellulosic Chemistry*, Second Edition, Revised, and Expanded, CRC Press, New York. ISBN: 0-8247-0024-4.

Jane, F. W. (1970). *Structure of Wood*, A & C Black Publishers Ltd., London. ISBN 10: 0713609125; ISBN 13: 9780713609127.

Lawoko, M., G. Henriksson and G. Gellerstedt (2006). Characterization of lignincarbohydrate complexes (LCCs) of spruce wood (*Picea Abies* L) isolated with two methods. *Holzforschung*, 60:156–61.

Smith, I., E. Landis and M. Gong (2003). *Fracture and Fatigue in Wood*. John Wiley & Sons. ISBN: 0-471-48708-2.

Tsoumis, G. (1991). *Science and Technology of Wood: Structure, Properties, Utilization*, Van Nostrand Reinhold, New York. ISBN: 0-442-23985-8.

Umney, N. and S. Rivers (2003). *Conservation of Furniture*, Butterworth Heinemann, Oxford. ISBN: 0-7506-09583.

Walker, J. C. F., B. G. Butterfield, J. M. Harris, T. A. G. Langrish and J. M. Uprichard (1993). *Primary Wood Processing: Principals and Practice*, Springer. ISBN: 978-94-015-8112-7.

2

Wood Mechanical Behaviour

2.1 Elastic and Strength Properties

Wood is an anisotropic material, i.e. its mechanical properties depend on the considered direction, and its mechanical behaviour is strongly affected by the complex material anatomy. In fact, wood is a natural biological material designed to resist loads (gravity, wind, inclination of terrains and others), by means of an internal structure that maximise strength and stiffness in stressed directions. Actually, wood cells are oblong and predominantly oriented in the grain direction, giving rise to a strong direction along the longitudinal axis of the trunk, while in the other two directions stiffness and strength are remarkably inferior. Accordingly, from a macroscopic point of view commonly assumed for structural design purposes, wood is generally considered as a cylindrical orthotropic material, with the principal axes of orthotropy (L, R, T) given by the longitudinal or grain direction of the tree trunk, radial and tangential directions (Figure 2.1).

There are large differences in stiffness and strength between these directions. In fact, the properties on the L direction, usually denoted as properties parallel to the grain due to direction of wood cellular structure, are substantially higher relative to the ones associated with the R and T axes. In these directions, the cellular structure is compressed in its weakest directions revealing similar properties values, currently designated as properties perpendicular to the grain. As a result, wood splitting along planes parallel to the fibres is the typical failure mode of this material, being often induced by tension perpendicular to the grain or shear loading.

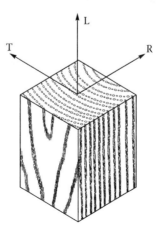

FIGURE 2.1
Orthotropic directions in wood.

The earlier considerations highlight the relevance of knowledge of wood elastic properties in the three directions (L, R, T). Following the orthotropic elasticity theory, the relation between strains (ε) and stresses (σ) can be expressed by the generalised Hooke's law

$$\varepsilon = \mathbf{S}\sigma \tag{2.1}$$

which gives

$$
\begin{Bmatrix} \varepsilon_L \\ \varepsilon_R \\ \varepsilon_T \\ \varepsilon_{RT} \\ \varepsilon_{LT} \\ \varepsilon_{LR} \end{Bmatrix} =
\begin{bmatrix}
\dfrac{1}{E_L} & \dfrac{-v_{RL}}{E_R} & \dfrac{-v_{TL}}{E_T} & 0 & 0 & 0 \\
\dfrac{-v_{LR}}{E_L} & \dfrac{1}{E_R} & \dfrac{-v_{TR}}{E_T} & 0 & 0 & 0 \\
\dfrac{-v_{LT}}{E_L} & \dfrac{-v_{RT}}{E_R} & \dfrac{1}{E_T} & 0 & 0 & 0 \\
0 & 0 & 0 & \dfrac{1}{G_{RT}} & 0 & 0 \\
0 & 0 & 0 & 0 & \dfrac{1}{G_{LT}} & 0 \\
0 & 0 & 0 & 0 & 0 & \dfrac{1}{G_{LR}}
\end{bmatrix}
\begin{Bmatrix} \sigma_L \\ \sigma_R \\ \sigma_T \\ \sigma_{RT} \\ \sigma_{LT} \\ \sigma_{LR} \end{Bmatrix} \tag{2.2}
$$

Using the concept of strain energy density, it can be proved that the compliance matrix ($\mathbf{S}$) is symmetric, which means that

$$\frac{v_{LR}}{E_L} = \frac{v_{RL}}{E_R}, \frac{v_{LT}}{E_L} = \frac{v_{TL}}{E_T}, \frac{v_{RT}}{E_R} = \frac{v_{TR}}{E_T} \tag{2.3}$$

As a result, the number of independent constants required to characterise the elastic behaviour of wood reduces to nine: the Young's moduli (E_L, E_T, E_R), the shear moduli (G_{TR}, G_{LR}, G_{LT}) and the Poisson's ratios (v_{LT}, v_{LR}, v_{TR}), which should be measured experimentally. The experimental tests are mostly established by the EN 408:2010+A1:2012 standard, which defines the requirements to measure fundamental physical properties of clear wood. This concept applies to material presenting the wood grain predominantly aligned, without knots of significant size, resin pockets or defects originated by fracture or fungi attack. Wood used in experimental tests should be stored in a normalised atmosphere of 20°C ± 2°C and 65% ± 5% of relative humidity till equilibrium is reached. Also, the specimen dimensions (i.e. height, h; width, B; and length, L) have to present a clearance of 1%, and the loading equipment used in the experiments should be able to measure the load with an accuracy of 1% of the load applied to the test piece.

2.1.1 Young's Moduli and Normal Strengths

The Young's moduli can be measured in tensile or compression tests considering specimens oriented according to the required direction (EN 408:2010+A1:2012). The determination of the longitudinal Young modulus E_L by tensile tests requires the preparation of the specimen geometry shown in Figure 2.2a. To avoid sliding relative to the machine grips in the course of the loading phase, specimen extremities have to be reinforced with steel (or wood) plates bonded with epoxy adhesive. The specimen must be fixed to machine grips

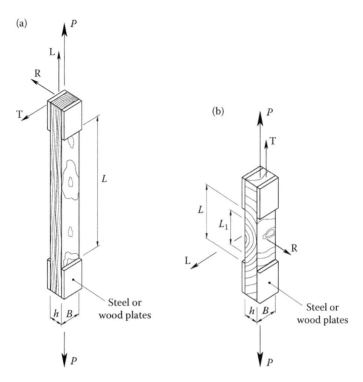

FIGURE 2.2
Specimen geometries and dimensions are used in the tensile tests to determine (a) E_L ($h = 45$, $B = 70$ mm and $L \geq 9B$) and (b) E_T ($h = 45$, $B = 70$ mm, $L = 180$ mm and $L_1 \approx 0.6L$).

in a way that traction is induced without bending. The load has to be applied at a constant strain rate not greater than 0.00005/s and rupture should be attained within 300 ± 120 s. A gauge length has to be considered over a length of five times the width of the specimen located not closer to the ends of the grips than twice the test piece width. Two extensometers must be used and positioned to minimise the effects of distortion. A similar procedure is followed to determine the Young moduli in directions perpendicular to grain (E_T or E_R) using shorter specimens and gauge length (L_1 in Figure 2.2b).

Compressive tests can also be used to determine Young moduli. Specimens have to be loaded concentrically between parallel metal plates (Figure 2.3) using spherically seated loading heads or other procedures, which assure compressive loading without bending. To measure the longitudinal modulus, the test piece is equipped with a gauge over a length of four times the width of the piece (i.e. $L_1 = 4B$, Figure 2.3a), located in the central part of the specimen; in the case of the transverse or radial moduli, this length is smaller (Figure 2.3b). The use of two extensometers is recommended to minimise eventual effects of distortion. An initial preload is applied to lock the loading heads that prevent the angular movement of the piece. The crosshead displacement rate may be lesser than (0.00005 L) mm/s.

The initial slope of the load–displacement curve ($P - \delta$) corrected for the cross-sectional area of the specimen ($A = Bh$) and distance covered by the used extensometer (L_1) provides the relation

$$E_i = \frac{\Delta P L_1}{\Delta \delta A} ; i = L, R, T \tag{2.4}$$

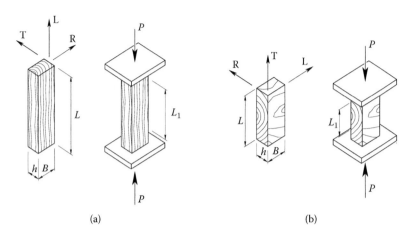

FIGURE 2.3
Specimen geometries used in the compression tests to determine (a) E_L ($h = 45$, $B = 70$ mm, $L = 6B$ and $L_1 = 4B$) and (b) E_T ($h = 45$, $B = 70$ mm, $L = 90$ mm and $L_1 \approx 0.6L$).

where ΔP and $\Delta \delta$ represent the increment of load and displacement, respectively, between two points located in the initial linear region of the load–displacement curve. The elastic moduli may also be obtained through a linear regression analysis over the load–displacement curve in the range of elastic deformation. In that case, the coefficient of determination has to be higher than 0.99.

Tensile and compression strengths in directions parallel and perpendicular to grain can also be determined using the same specimen shapes, dimensions and experimental setup utilised for elastic moduli (Figure 2.3). The corresponding tensile (subscript t) or compressive (subscript c) strength is given by

$$\sigma_{ui,j} = \frac{P_{max}}{A} ; i = L, R, T; j = t, c \tag{2.5}$$

The mechanical test is performed under displacement control by programming the crosshead velocity to induce the rupture of the specimen within 300 ± 120 s.

Alternately, bending tests can be used for measuring the Young's moduli in the fibre direction (EN 408:2010+A1:2012). In the case of the three-point bending test, the following equation should be used

$$E_L = \frac{\Delta P L^3}{\Delta \delta 48 I} \tag{2.6}$$

with L being the specimen length, I the second moment of area of the cross section of the specimen (i.e. $Bh^3/12$), and ΔP and $\Delta \delta$ the increment of load and deflection at the centre of the beam in the elastic regime. The modulus given by Eq. (2.6) is known as 'apparent modulus', since shear effects existing in three-point bending tests are not accounted for. The agreement between the apparent modulus with the true one increases with the increase of span-to-depth and Young versus shear modulus ratios. The ASTM D3737 standard points to span-to-depth ratio of 21 and Young versus shear modulus ratio of 14, to achieve an apparent modulus equal to 95% of the true one. When such conditions are not fulfilled, the four-point bending test characterised for pure bending between the loading points (Figure 2.4) should be used. The beam is simply supported and symmetrically loaded. Small steel plates ($w \leq h/2$) are disposed in the supports and loading points to prevent wood indentation during loading. A lightweight rigid frame is fixed to each side of the beam in two points at the half-height (i.e. $h/2$). This procedure allows attaching the device at the half span of the beam to provide the displacement measurement δ (mean value of each acquisition) relative to the neutral axis, by means of two gauges ($L_1 = 5h$). Lateral restraint with minimal frictional resistance on the beam may be necessary to prevent specimen buckling during loading. Loading is applied under displacement control (not greater than $0.003h$ mm/s) without exceeding the proportional limit or damage onset. The expression for the local Young's moduli becomes (EN 408:2010+A1:2012)

$$E_{L,local} = \frac{\Delta P s L_1^2}{\Delta \delta 16 I} \tag{2.7}$$

with s being the distance between a loading point and the nearest support. This test also provides the bending strength

$$\sigma_{ui,j} = \frac{3 s P_{max}}{B h^2} \tag{2.8}$$

A similar arrangement can be used to measure the global modulus of elasticity in bending (Figure 2.5). The global modulus in the fibre's direction is calculated as follows:

$$E_{L,global} = \frac{3 s L^2 - 4 s^3}{2 B h^3 \left(2 \dfrac{\Delta \delta}{\Delta P} - \dfrac{6 s}{5 G_{Li} B h} \right)} ; i = R, T \tag{2.9}$$

The variations of load and applied displacement in this equation (ΔP and $\Delta \delta$) are obtained from the initial linear load–displacement relationship, including at least the loading values of $0.2 F_{max,est}$ and $0.3 F_{max,est}$. The coefficient of determination (R^2) should be equal to 0.99, at least. Equation (2.9) accounts for shear deformation, which means that the shear modulus G_{Li} has to be determined (Section 2.1.3).

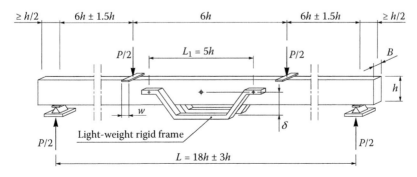

FIGURE 2.4
Test arrangement to measure the local modulus of elasticity in four-point bending.

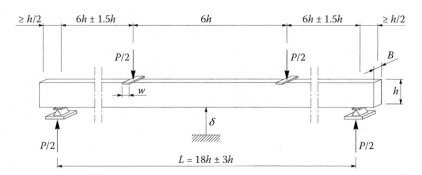

FIGURE 2.5
Test arrangement to measure the global modulus of elasticity in bending.

2.1.2 Poisson's Ratios

Poisson's ratios quantify the lateral contraction resulting from a longitudinal load. They may be evaluated in tensile tests considering two strain gauges bonded to the specimen and oriented in the axial and transversal directions to measure the strains in those directions (Figure 2.6), thus leading to

$$\nu_{ij} = -\frac{\varepsilon_j}{\varepsilon_i}; \quad i,j = L,R,T \text{ and } i \neq j \tag{2.10}$$

The strain measurements must be performed in the elastic regime.

2.1.3 Shear Moduli and Strengths

The determination of shear moduli is more complex, and there is some controversy about the appropriate procedure. The EN 408:2010+A1:2012 proposes two techniques. In the first one, the shear modulus is calculated from the Young's moduli determined by three-point ($E_{i(TPB)}$ from Eq. 2.6) and four-point ($E_{i(FPB)}$ from Eq. 2.7) bending tests assuming in the former $L = L_1 = 5h$. The shear modulus is then given by the following equation:

$$G_{ij} = \frac{k_G h^2}{L_1^2 \left(\frac{1}{E_{i(TPB)}} - \frac{1}{E_{i(FPB)}} \right)}; \quad ij = RT, LT, LR \tag{2.11}$$

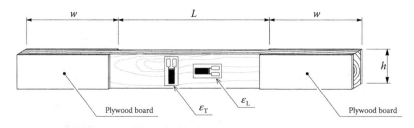

FIGURE 2.6
Schematic representation of the strain gauges disposition on a tangential-longitudinal (TL) specimen used to determine the ν_{LT} Poisson's ratio ($L = 90$, $w = 55$ in mm).

where $k_G = 1.2$ for rectangular or square cross sections. The second procedure consists of loading a specimen in three-point bending, considering, at least, four different spans chosen to provide approximately equal increments of $(h/L)^2$ between them within the range of 0.0025–0.035. The slope (K_1) of the plot of $1/E_i$ versus $(h/L)^2$ is used to determine the shear modulus

$$G_{ij} = \frac{k_G}{K_1}; \quad ij = RT, LT, LR \qquad (2.12)$$

The described methods present some limitations. They require specimens with a certain length that are not easy to obtain in the RT plane. On the other hand, some authors (Yoshihara et al., 1998) have pointed that those methods give rise to shear modulus values an order of magnitude lower than expected.

The recent version of the standard EN 408:2010+A1:2012 proposes two distinct experimental methods to measure the shear modulus of wood. The first one is based on the torsional test of a wooden beam, while the second presents a bending test to create a shear field. The former is more adequate to massive wood, while the latter is indicated to glued laminated timber (glulam). Both methods require a wood beam with a constant rectangular shape (height: h; width: B) and length greater than $19h$.

The torsional test requires a wooden beam firmly constrained in two supports (span of $16h$), as shown in Figure 2.7. According to this experimental setup, a shear moment T_r is applied to the wood beam by means of a ring rigidly fixed to one of the mentioned supports (right side in Figure 2.7). Once loaded in shear, two angular displacements φ_1 and φ_2 are recorded in sections (1) and (2) of the specimen (Figure 2.7), respectively, at a distance L_1 from each other. The rate of movement of the shear load is then obtained as follows:

$$\frac{d\varphi}{dt} = \frac{\sigma_{ij}\chi}{225 G_{ij} \eta}\left(\frac{L_1}{h}\right); \quad ij = LT, LR \qquad (2.13)$$

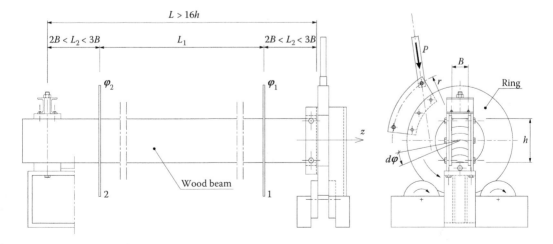

FIGURE 2.7
Test arrangement to measure the shear modulus.

being σ_{ij} the wood characteristic shear strength parallel to the grain (in MPa), χ and η shape coefficients (Table 2.1), G_{ij} the shear modulus of the beam and L_1 the distance between sections (1) and (2).

The relation between the torsional applied moment T_r and the relative shear displacement φ, represented by the torsional stiffness k_{tor}, is evaluated by means of the calculated linear regression of T_r as a function of the recorded values of φ (Figure 2.8) obtained in the experiments within the range of elastic deformation. The referred linear regression should exhibit, at least, a coefficient of determination (R^2) of 0.98.

In the experiments, the maximum value of the applied shear moment T_r should be reached in less than 150 sec and cannot exceed the limit of proportionality of wood or induce damage in the segment (1)–(2) of the beam (Figure 2.7). For this reason, a threshold for this moment is defined as follows:

$$T_r = \frac{2}{3} B^2 h \sigma_{ij} \chi \qquad (2.14)$$

The wood shear modulus is measured through

$$G_{ij} = \frac{k_{tor}}{\eta h B^3} L_1; \quad ij = LT, LR \qquad (2.15)$$

Owing to the anisotropic nature of wood, the twist of a wood specimen in a torsion test involves two shear moduli, i.e. G_{LT} and G_{LR}, which constitutes a disadvantage of this test. To have a good estimate of one of these properties, it is advisable to consider larger ratios h/B so that torsion can be primarily dependent upon one of the shear moduli.

The standard EN 408:2010+A1:2012 also proposes the use of the standardised four-point bending test (Figure 2.4) to measure the shear modulus. Due to minor induced

TABLE 2.1
Shape Coefficients of the Shear Test according to EN 408:2010+A1:2012 Standard

h/B	1.0	1.2	1.5	2.0	2.5	3	4	5	10
η	0.140 6	0.166	0.196	0.229	0.249	0.263	0.281	0.291	0.312
χ	0.415 8	0.456 4	0.461 8	0.490 4	0.516 2	0.533 4	0.563 4	0.596 0	0.627 0

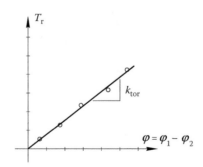

FIGURE 2.8
Linear regression of the applied torsional moment and relative shear displacement.

shear deflection, a rigorous measurement device should be applied. In this context, four extensometers are installed on the left–right, front–back sides of the beam in the segment where the transverse load is constant [i.e. $(6h \pm 1.5h)/2$ from the support] at the mid-height (Figure 2.9). These extensometers should form two diagonals of the same length (h_0), which increase as the beam is loaded during bending becoming equal to $h_0 \pm \delta_{i,j}$, where i refers the specimen side and j refers to the diagonal (Figure 2.9). The transversal shear modulus is given by

$$G_{Li} = \alpha \frac{h_0}{Bh} \frac{\Delta V}{\Delta \delta}; \quad i = R, T \qquad (2.16)$$

where ΔV represents the increment of shear force within the elastic range and $\Delta \delta$ the corresponding average deformation measured in the extensometers. The parameter α accounts for the distribution of shear stress over the whole depth of the beam and is given by

$$\alpha = \frac{3}{2} - \frac{h_0^2}{4h^2} \qquad (2.17)$$

Some alternative methods have also been proposed to characterise shear behaviour of wood. The off-axis tensile, the Iosipescu and the Arcan tests have been applied recently in wood for this purpose and are succinctly described in the following.

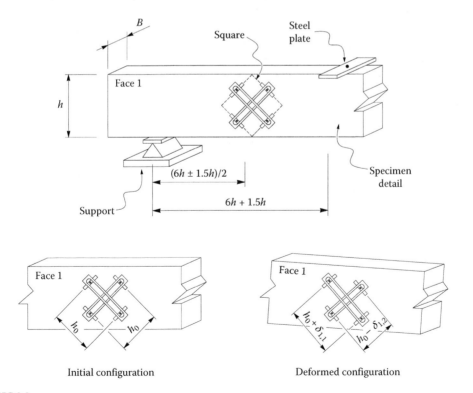

FIGURE 2.9
Detail of the bending test presented in Figure 2.4, showing the square arrangement of extensometers fixed to a face of the beam, both in the initial and deformed configurations.

The off-axis test (Xavier et al., 2004) consists of a tensile test on a parallelepiped specimen in which one of the material symmetric directions is rotated (angle α) relative to the longitudinal direction of the specimen (Figure 2.10).

The shear strain is measured using a three-element strain gauge rosette (a, b, c) bonded at the specimen centre. The normal strain readings at the 60° delta rosette (ε_a, ε_b, ε_c) are used to get the shear strain in the material principal axes

$$\varepsilon_{ij} = (\varepsilon_a - 2\varepsilon_b + \varepsilon_c)\sin\alpha + (\varepsilon_a - \varepsilon_c)\cos\alpha; \quad ij = RT, LT, LR \qquad (2.18)$$

The stresses in the material principal axis (i, j) can be found by stress transformation equations as follows:

$$\sigma_i = \frac{P}{A}\cos^2\alpha; \; \sigma_j = \frac{P}{A}\sin^2\alpha; \; \sigma_{ij} = -\frac{P}{A}\sin\alpha\cos\alpha; \quad ij = RT, LT, LR \qquad (2.19)$$

The shear modulus can be obtained from the ratio between shear stress and strain components

$$G_{ij} = \frac{\sigma_{ij}}{\varepsilon_{ij}}; \quad ij = RT, LT, LR \qquad (2.20)$$

Xavier et al. (2004) have found $\alpha = 15°$ (Figure 2.10) as being the angle that maximises the ratio $\varepsilon_{ij}/\varepsilon_x$ in the case of *Pinus Pinaster* Ait. species.

The off-axis tensile test can also be used to estimate the shear strength of wood in the referred material planes. However, a direct identification is not viable since a pure shear stress state is not achieved in the off-axis tensile test, thus being necessary to apply a suitable interactive strength criterion (for example, the Tsai–Hill or Tsai–Wu criteria) to identify this property. This will be discussed later in the next section.

In the Iosipescu test, the specimen is a rectangular beam of small dimensions with symmetric V-notches at its centre (Figure 2.11). An appropriate setup should be used for the Iosipescu test. The specimen is clamped in one of its extremities and loaded at the other one by displacement of the machine support to induce shear loading at the specimen's minimum cross-section area. The design of the V-notches aims to induce a quasi-uniform shear stress distribution at the critical section, whose average is given by

$$\sigma_{ij}^{av} = \frac{P}{A}; \quad ij = RT, LT, LR \qquad (2.21)$$

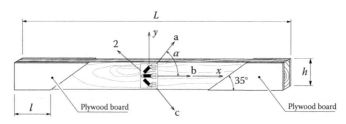

FIGURE 2.10
Off-axis tensile test [$L = 200$, $l = 27$, $h = 20$ (mm), $\alpha = 60°$].

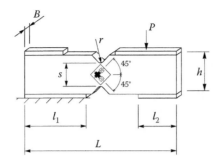

FIGURE 2.11
Specimen configuration for the Iosipescu test $L = 80$, $l_1 = 32$, $l_2 = 20$, $h = 20$, $s = 12$, $r = 2$, $B = 8$ (mm).

The shear strain is measured by means of a strain gauge rosette, with two elements oriented at ±45° relative to longitudinal axis. The shear strain in the material principal axis becomes

$$\varepsilon_{ij}^{av} = \varepsilon_{45°} - \varepsilon_{-45°}; \quad ij = RT, LT, LR \tag{2.22}$$

thus allowing the determination of the apparent shear modulus using Eq. (2.20). The conditions of uniform stress and strain are not completely satisfied in anisotropic materials. Xavier et al. (2004) proposed the consideration of two correction factors determined by finite element analysis to account for the non-uniformity of stress distribution (C) and for the difference between the shear strain at the central point and the gauge reading (S)

$$G_{ij} = CSG_{ij}^{ap}; \quad ij = RT, LT, LR \tag{2.23}$$

The shear strength can be calculated from Eq. (2.21) considering the load corresponding to specimen failure. However, since a pure shear stress state does not occur over the gauge length region, it is advisable to perform finite element analysis to determine the stress field. Subsequently, an interactive failure criterion should be used to take into account all stress components and obtain a valid material shear strength.

Shear characterisation of wood can also be carried out by the Arcan test (Xavier et al., 2009). The specimen geometry is similar to the one utilised in the Iosipescu test (Figure 2.12a), although the loading mode is different. The specimen is mounted on a

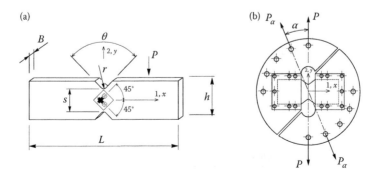

FIGURE 2.12
Schematic representation of the (a) Arcan specimen and (b) test setup. $L = 80$, $h = 20$, $s = 12$, $r = 2$, $B = 8$ (mm).

26 *Wood Fracture Characterisation*

special device, which allows loading to be applied under several directions (angle α in Figure 2.12b). An almost pure shear loading is obtained in the minimum cross section of the specimen for $\alpha = 0°$. Owing to similarity with the Iosipescu test, the same procedure can be followed to determine the shear moduli and shear strengths of wood in the selected three planes. As a result, Eqs. (2.18–2.21) apply directly to obtain those properties.

The wood species used in the work described in this book is the *Pinus pinaster* Ait. The elastic and strength properties were measured by means of the tests and procedures described earlier and are presented in Tables 2.2 and 2.3.

2.2 Strength Failure Criteria

Structural applications of wood are generally under multiaxial loading, which leads to complex stress state. To predict wood failure under these general situations, it is necessary to apply appropriate failure criteria. Since wood is viewed as an orthotropic material, the criteria developed for other orthotropic materials, as is the case of artificial composites, can be employed.

Two types of criteria can be utilised to identify wood failure: non-interactive and interactive. Non-interactive criteria do not account for combined effects of stress or strain components, instead, failure is achieved when one of the stress or strain components reaches the corresponding strength value. The maximum stress and maximum strain criteria belong to this group. The maximum stress criterion predicts failure when one of the stress tensor in the material symmetry system reaches the corresponding strength,

$$\sigma_i = \sigma_{ui}^t \text{ for } \sigma_i > 0; \quad i = L, R, T$$
$$\sigma_i = \sigma_{ui}^c \text{ for } \sigma_i < 0; \quad i = L, R, T \tag{2.24}$$
$$|\sigma_{ij}| = \sigma_{uij}; \quad ij = RT, LT, LR$$

where superscripts 't' and 'c' denote the tensile and compressive strengths, respectively. The maximum strain criterion considers that failure occurs when one of the strain components in the material symmetry system attains the corresponding failure value, thus yielding

TABLE 2.2

Nominal Elastic Properties of *Pinus pinaster* Ait

E_L (GPa)	E_R (GPa)	E_T (GPa)	ν_{LT}	ν_{LR}	ν_{TR}	G_{LR} (GPa)	G_{LT} (GPa)	G_{TR} (GPa)
12.0	1.91	1.01	0.51	0.47	0.31	1.12	1.04	0.29

TABLE 2.3

Nominal Strength Properties of *Pinus pinaster* Ait

σ_{uL} (MPa)	σ_{uR} (MPa)	σ_{uT} (MPa)	σ_{uLR} (MPa)	σ_{uLT} (MPa)	σ_{uRT} (MPa)
97.5	7.9	4.2	16.0	16.0	4.5

Wood Mechanical Behaviour 27

$$\varepsilon_i = \varepsilon_{ui}^t \text{ for } \varepsilon_i > 0; \quad i = L,R,T$$
$$\varepsilon_i = \varepsilon_{ui}^c \text{ for } \varepsilon_i < 0; \quad i = L,R,T \quad (2.25)$$
$$|\varepsilon_{ij}| = \varepsilon_{uij}; \quad ij = RT,LT,LR$$

This criterion can be expressed in terms of stresses using the constitutive equations and produces similar results when compared with the maximum stress criterion. These non-interactive criteria present the advantage of immediate identification of failure mode, but they are non-conservative since they do not include the combined effects of all stress/strain components resulting from applied loading.

The interactive criteria are usually based on stress quadratic functions. One of the pioneering interactive criterion for wood is the empirical Hankinson (1921) criterion widely used in industry. This formula predicts the strength of wood under uniaxial compressive loading not coincident with the grain orientation. Considering, for example, the LR plane,

$$\sigma_\theta = -\frac{\sigma_{uL}^c \sigma_{uR}^c}{\sigma_{uL}^c \sin^2 \theta + \sigma_{uR}^c \cos \theta} \quad (2.26)$$

with θ being the angle between grain orientation and applied loading. A similar expression can also be employed to predict wood strength under uniaxial tensile loading in a direction not coincident with grain orientation (Kollmann and Coté, 1984)

$$\sigma_\theta = \frac{\sigma_{uL}^t \sigma_{uR}^t}{\sigma_{uL}^t \sin^n \theta + \sigma_{uR}^t \cos^n \theta} \quad (2.27)$$

where n is a constant ranging between 1.5 and 2.0. Although strictly empirical, the Hankinson formulas have presented values in agreement with the experiments. However, the application of wood in complex structures requires the use of more general criteria appropriate for orthotropic materials. One of these cases is the Tsai–Hill criterion that derives from the Hill expression developed for metals. For the general case of tridimensional loading for orthotropic materials, the Tsai–Hill criterion (Tsai, 1968) establishes

$$A(\sigma_R - \sigma_T)^2 + B(\sigma_T - \sigma_L)^2 + C(\sigma_L - \sigma_R)^2 + 2D\sigma_{RT}^2 + 2E\sigma_{LT}^2 + 2F\sigma_{LR}^2 = 1 \quad (2.28)$$

where $A, B, ..., F$ are constants depending on failure strengths

$$B + C = \frac{1}{\sigma_{uL}^2}; \quad A + C = \frac{1}{\sigma_{uR}^2}; \quad A + B = \frac{1}{\sigma_{uT}^2};$$
$$2D = \frac{1}{\sigma_{uRT}^2}; \quad 2E = \frac{1}{\sigma_{uLT}^2}; \quad 2F = \frac{1}{\sigma_{uLR}^2}. \quad (2.29)$$

As an example, in the case of plane stress in the LR plane, the Tsai–Hill criterion predicts failure when

$$\frac{\sigma_L^2}{\sigma_{uL}^2} - \frac{\sigma_L \sigma_R}{\sigma_{uL}^2} + \frac{\sigma_R^2}{\sigma_{uR}^2} + \frac{\sigma_{LR}^2}{\sigma_{uLR}^2} = 1. \quad (2.30)$$

Norris (1962) proposed a similar criterion with the difference of considering different strengths for tensile and compressive loading.

The earlier described expressions derive from criteria developed for isotropic materials. Consequently, some assumptions do not verify anisotropic materials. Specifically, states of hydrostatic stresses do not induce plastic yielding in metals that justifies the exclusive use of quadratic terms. However, as in the case of other composite materials, wood can fail under hydrostatic stress state. To account for this difference, Tsai and Wu (1971) proposed a tensorial quadratic polynomial containing linear terms

$$F_{ij}\sigma_{ij} + F_{ijkl}\sigma_{ij}\sigma_{kl} = 1; \quad i,j,k,l = \mathrm{L,R,T} \tag{2.31}$$

where F_i and F_{ij} are strength tensors determined experimentally. In the case of orthotropic materials under plane stress in the LR plane, the Tsai–Wu criterion yields to

$$\left(\frac{1}{\sigma_{uL}^t} - \frac{1}{\sigma_{uL}^c}\right)\sigma_L + \left(\frac{1}{\sigma_{uR}^t} - \frac{1}{\sigma_{uR}^c}\right)\sigma_R + \frac{\sigma_L^2}{\sigma_{uL}^t\sigma_{uL}^c} + \frac{\sigma_R^2}{\sigma_{uR}^t\sigma_{uR}^c} + 2F_{LR}\sigma_L\sigma_R + \frac{\sigma_{LR}}{\sigma_{uLR}} = 1 \tag{2.32}$$

The quadratic terms define an ellipsoid in the stress space and account for the interactions between normal stresses. The linear terms allow to account for different tensile and compressive strengths that makes this criterion more consistent with the real behaviour of composites such as wood. The required strengths can be measured by uniaxial tests. The only parameter left to be determined is the interaction stress coefficient F_{LR}. The rigorous measurement of this coefficient would require biaxial loading tests with stresses σ_L and σ_R, that are very difficult to perform in wood. Moreover, the material inhomogeneity leads to remarkable data scatter, giving rise to a considerable amount of uncertainty on F_{LR}. To overcome these drawbacks, Liu (1984) combined the results of an off-axis test with the Hankinson's formula to obtain

$$F_{LR} = \frac{1}{2}\left(\frac{1}{\sigma_{uL}^t\sigma_{uR}^c} + \frac{1}{\sigma_{uL}^c\sigma_{uR}^t} - \frac{1}{\sigma_{uLR}^2}\right)^{1/2} \tag{2.33}$$

This procedure satisfied the stability condition

$$F_{LL}F_{RR} - F_{LR}^2 \geq 0 \tag{2.34}$$

that must be respected for the criterion equation to represent a closed surface. In addition, the proposed procedure replaces the execution of the cumbersome biaxial test by a simple off-axis test.

The earlier described non-interactive and interactive strength criteria are stress or strain based. Many structural applications of wood involve discontinuities and singularities such as notches or holes, leading to important stress concentration effects. Additionally, wood as a natural and biological material presents drastic variations in its inner structure because of internal defects such as knots, variation of grain orientation, reaction wood and others. In addition, in applications exposed to outdoor conditions, wood becomes susceptible to degradation due to a variety of natural causes. It is also susceptible to rot, insect attack and fungi decay, although appropriate preservative treatments can avoid or delay these unwanted effects. It can then be concluded that there is a considerable source of variability at several levels. The consequence is the consideration of several high safety factors in the design of wood structures. As an example, the *'Eurocode 5: Design of timber structures'* preconises the use of four safety factors in the design of the simple case of a wood cantilever beam loaded at its extremity.

To overcome these drawbacks and limitations, a promising line of research consists in the application of fracture mechanics concepts in wood design. Such methodology can contribute significantly to a better understanding and more reliable design methods concerning the project of wood structural applications. A comprehensive description of the fundaments of fracture mechanics science as well as the necessary extensions to account for wood specificities are presented in the following section.

2.3 Fracture Mechanics Based Approaches

2.3.1 Linear Elastic Fracture Mechanics

Two main approaches can be used to predict damage in a structure. One of them is based on the strength of materials concepts where materials are assumed free of defects. However, the application of the strength of materials-based criteria in structural design presents some drawbacks. In fact, in many situations, the problem of stress concentrations nearby to a notch or a flaw leads to mesh dependency in numerical approaches. In these cases, an accurate prediction is not possible using finite element analysis and considering such criteria. Unlike to what happens in these methodologies, fracture mechanics based approaches (Anderson, 2005) assume the presence of an inherent defect in the material. This initial defect (or crack) can propagate under three different loading modes, as can be seen in Figure 2.13. Mode I represents an open mode and the others (mode II and mode III) are shear modes. In the majority of real situations, the applied loading originates a combination of modes at the crack tip, which implies that a mixed-mode criterion should be considered to better simulate the damage propagation.

There are two types of fracture mechanics criteria based on stress intensity factors or on the energetic concepts. The stress intensity factor is defined as

$$K = Y\sigma_R\sqrt{\pi a} \tag{2.35}$$

where Y is a non-dimensional factor depending on the geometry and loading distribution, σ_R is the remote applied stress and a is the crack length. It is assumed that crack propagation occurs when the stress intensity factor attains a critical value

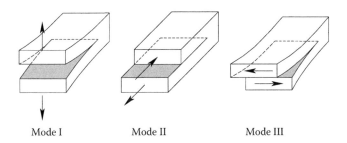

FIGURE 2.13
Loading modes.

$$K_c = \sigma_u \sqrt{\pi a} \tag{2.36}$$

where σ_u is the strength of the material.

The energetic criterion is based on the assumption that crack growth will occur when the energy available at the crack tip (G, strain energy release rate) induced by the applied loading overcomes the critical strain energy release rate (G_c), which is a material property. The strain energy release rate is given by

$$G = \frac{dW}{dA} - \frac{dU}{dA} \tag{2.37}$$

where W is the work performed by external forces, U the internal strain energy and dA the variation of crack surface. Considering a general body with constant thickness B (Figure 2.14) under loading P normal to crack plane, it can be written

$$W = P\delta; \quad U = \frac{1}{2}P\delta \tag{2.38}$$

Combining Eqs. (2.37) and (2.38) and using the body compliance $C = \delta/P$ yields

$$G = \frac{P^2}{2B} \frac{dC}{da} \tag{2.39}$$

This is the Irwin–Kies relation (Irwin and Kies, 1954), considered as one of the fundamental equations of linear elastic fracture mechanics (LEFM). During propagation, the values of G given by previous equation define the energy necessary to crack growth (G_c), thus characterising material fracture behaviour.

It should be noted that G and K are intrinsically related. In fact, Irwin (1957) demonstrated that under plane stress conditions

$$G = \frac{K^2}{E} \tag{2.40}$$

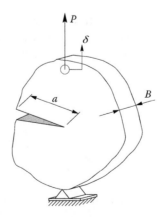

FIGURE 2.14
Cracked body under uniaxial loading.

and under plane strain conditions

$$G = \frac{K^2(1-v^2)}{E} \tag{2.41}$$

where E and v are the Young's modulus and Poisson's ratio, respectively. These relationships are also valid for the respective critical values (G_c and K_c).

Since wood is a heterogeneous material, the energetic methods of fracture are more appropriate. In fact, the stress intensity factor is a local parameter, which can be affected by material heterogeneity. On the other hand, the energetic quantities constitute global fracture parameters, giving a more accurate idea of the fracture process in the material and are not drastically affected by local material variations. Actually, the strain energy release rate has an important physical significance related to the energy absorption during fracture.

One of the most popular energetic methods based on fracture mechanics concepts is the virtual crack closure technique, which is detailed by Krueger (2002). The strain energy release rates are obtained, assuming that the energy released when crack grows is equal to the work necessary to close it to its initial length before propagation. Considering a two-dimensional problem (see Figure 2.15), the strain energies (G_I and G_{II}) can be calculated by the product of the relative displacements at the 'opened point' (nodes l_1 and l_2) and the loads at the 'closed point' (node i)

$$\begin{aligned} G_I &= \frac{1}{2B\Delta a} Y_i \Delta v_1 \\ G_{II} &= \frac{1}{2B\Delta a} X_i \Delta u_1 \end{aligned} \tag{2.42}$$

with $B\Delta a$ being the area of the new surface created by an increment of crack propagation (see Figure 2.15). It should be assured that self-similar propagation occurs and an adequate refined mesh is used. The strain energy release rate components can be used in an energetic general power-law mixed-mode criterion

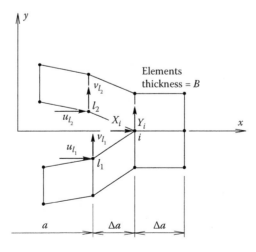

FIGURE 2.15
The virtual crack closure technique.

$$\left(\frac{G_\mathrm{I}}{G_\mathrm{Ic}}\right)^\alpha + \left(\frac{G_\mathrm{II}}{G_\mathrm{IIc}}\right)^\beta = 1 \tag{2.43}$$

to simulate damage propagation. G_Ic and G_IIc represent the critical strain energy release rates in mode I and mode II, respectively, and (α, β) are mixed-mode fracture parameters determined from material tests.

The application of fracture mechanics concepts to wood is still in an initial stage and in the majority of cases is limited to LEFM. However, the fundaments of LEFM are not completely satisfied in wood. In fact, LEFM assumes that the fracture process zone at the crack tip (region where several inelastic and dissipative processes take place like micro-cracking and fibre-bridging) is negligible relative to the body thickness, crack length and remaining ligament length located in the crack path. However, the FPZ in wood is responsible for a non-negligible dissipation of energy, thus being classified as a quasi-brittle material. In fact, when a non-damaged wood component is loaded, small cracks develop from microscopic defects, leading to an increase of compliance that is visible in the profile of the load–displacement curve. The multiplication of these small cracks leads to the development of a macro-crack and an FPZ mainly constituted by micro-cracking and fibre-bridging. The gradual and self-similar crack growth reflects on a progressive reduction of load in the post-peak region of the load–displacement curve. This fracture mechanism implies that LEFM is not appropriate for wood fracture characterisation since the energy dissipated in the FPZ is non-negligible. On the other hand, LEFM does not allow simulating or predicting crack initiation as it assumes the existence of a pre-crack. This aspect is crucial since the formation of cracks in wood is frequent.

2.3.2 Cohesive Zone Models

As discussed in the previous section, the strength of materials and fracture mechanics based criteria presents some disadvantages. The former presents mesh dependency during numerical analysis due to stress singularities, while the latter requires the definition of a pre-crack and a negligible FPZ. However, the strength of materials-based analysis provides a clear identification of the critical points in a structure, thus being valid for crack initiation analysis. On the other hand, fracture mechanics based methods are adequate to simulate damage propagation.

To overcome the referred drawbacks and exploit the usefulness of the described advantages, cohesive zone model (CZM) emerges as a suitable option. These methods combine aspects of stress-based analysis to model damage initiation and fracture mechanics to deal with damage propagation. Thus, it is not necessary to take into consideration an initial defect, and mesh dependency problems are overcome. In addition, they are able to simulate damage onset (no initial crack is needed) and non-self-similar crack growth without user intervention. CZM are usually based on a softening relationship between stresses and relative displacements between crack faces, thus simulating a gradual degradation of material properties and accounting for the damaging processes typical of wood. The referred relationship is usually introduced by means of interface finite elements, with null thickness located at the planes where damage is prone to occur in a finite element analysis. In this book, CZMs are widely used to simulate wood fracture in specimens and in structural applications. Therefore, a mixed-mode I + II trapezoidal with bilinear softening cohesive law is presented in the following, owing to its generality. Overall, during pure or mixed-mode analyses that will be discussed in this book, simpler laws are used. As they are particular cases of the presented one, it is easy for the reader to obtain the corresponding equations.

2.3.2.1 Trapezoidal with Bilinear Softening Cohesive Law

The fundamental equations of the trapezoidal with bilinear softening cohesive law for mixed-mode I + II loading are presented in this section. The CZM is based on a constitutive relationship between stresses (σ) and relative displacements (δ) at the integration points of the cohesive elements. Before damage starts to develop,

$$\sigma = E\delta \tag{2.44}$$

where **E** is the matrix that contains the stiffness parameter k in its main diagonal. The value of k (i.e. $10^6 - 10^7$ N/mm³) is selected according to a compromise between two important aspects: low values do not impede interpenetration between the cohesive element faces; too high values lead to numerical instabilities (Gonçalves et al., 2000). After damage onset, the previous equation is altered to include a damage parameter (diagonal matrix **D**)

$$\sigma = (\mathbf{I} - \mathbf{D})\mathbf{E}\delta \tag{2.45}$$

where **I** is the identity matrix. Therefore, the main goal of the model is to define the evolution of the damage parameter during the fracture process. The quadratic stress criterion is used to identify damage onset under mixed-mode I + II loading,

$$\left(\frac{\sigma_{1m,I}}{\sigma_{1,I}}\right)^2 + \left(\frac{\sigma_{1m,II}}{\sigma_{1,II}}\right)^2 = 1 \quad \text{if } \sigma_I > 0$$

$$\sigma_{1m,II} = \sigma_{1,II} \quad \text{if } \sigma_I \leq 0 \tag{2.46}$$

where $\sigma_{1m,i}$ and $\sigma_{1,i}$ (i = I, II) represent, respectively, stress components in mixed-mode and strengths of each pure mode (Figure 2.16). Combining Eqs. (2.44) and (2.46) for $\sigma_I > 0$,

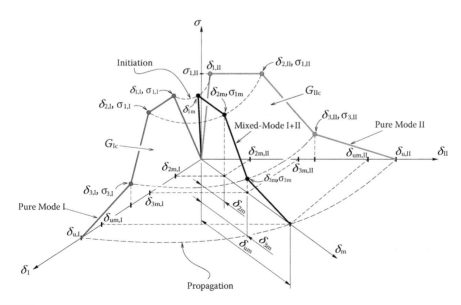

FIGURE 2.16
Trapezoidal with bilinear softening cohesive mixed-mode I + II zone model (de Moura et al., 2015).

$$\left(\frac{\delta_{1m,I}}{\delta_{1,I}}\right)^2 + \left(\frac{\delta_{1m,II}}{\delta_{1,II}}\right)^2 = 1 \tag{2.47}$$

where $\delta_{1,i}$ and $\delta_{1m,i}$ (i = I, II) are, respectively, the critical displacements corresponding to damage onset under pure and mixed-mode loading (Figure 2.16). Defining a mode ratio $\beta = |\delta_{II}|/\delta_I$ and the equivalent current mixed-mode displacement $\delta_m = \sqrt{\delta_{m,I}^2 + \delta_{m,II}^2}$, the equivalent displacement leading to damage onset can be defined as (Figure 2.16)

$$\delta_{1m} = \frac{\delta_{1,I}\delta_{1,II}\sqrt{1+\beta^2}}{\sqrt{\delta_{1,II}^2 + \beta^2\delta_{1,I}^2}} \tag{2.48}$$

The quadratic stress-based criteria (Eq. 2.46) and the equivalent one expressed in terms of relative displacements (Eq. 2.47) were also applied at the other inflection points (Figure 2.16), leading to the following relations:

$$\delta_{im} = \frac{\delta_{i,I}\delta_{i,II}\sqrt{1+\beta^2}}{\sqrt{\delta_{i,II}^2 + \beta^2\delta_{i,I}^2}} \quad (i = 2,3); \ \sigma_{3m} = \frac{\delta_{3,I}\delta_{3,II}\sqrt{1+\beta^2}}{\sqrt{\delta_{3,II}^2 + \beta^2\delta_{3,I}^2}} \tag{2.49}$$

where δ_{2m}, δ_{3m} and σ_{3m} represent, respectively, the equivalent relative displacements and stress in mixed-mode I + II at the inflection points. The linear energetic criterion was used to simulate damage propagation

$$\frac{G_I}{G_{Ic}} + \frac{G_{II}}{G_{IIc}} = 1 \tag{2.50}$$

where G_i, i = (I, II) is given by the area circumscribed by the trapezoids obtained by the projection of the mixed-mode trapezoid in the planes of mode I and II

$$G_i = \frac{1}{2}\left(\sigma_{1m,i}(\delta_{2m,i} + \delta_{3m,i} - \delta_{1m,i}) + \sigma_{3m,i}\left(\delta_{um,i} - \delta_{2m,i}\right)\right) \tag{2.51}$$

The ultimate relative displacement corresponding to complete failure (δ_{um}) is obtained substituting Eq. (2.51) in Eq. (2.50) and accounting for the mode ratio β and equivalent mixed-mode displacements defined earlier:

$$\delta_{um} = \frac{\dfrac{2G_{Ic}G_{IIc}\left(1+\beta^2\right)}{G_{IIc} + \beta^2 G_{Ic}} - k\delta_{1m}\left(\delta_{2m} + \delta_{3m} - \delta_{1m}\right)}{\sigma_{3m}} + \delta_{2m} \tag{2.52}$$

The equivalent displacements of the inflection points [Eqs. (2.48, 2.49 and 2.52)] are used to define the evolution of the damage parameter in the three branches of the softening law. In the plateau region (i.e. for $\delta_{1m} \leq \delta_m \leq \delta_{2m}$) the damage parameter is given by

$$d_m = 1 - \frac{\delta_{1m}}{\delta_m} \tag{2.53}$$

In the first descending softening branch (i.e. for $\delta_{2m} \leq \delta_m \leq \delta_{3m}$),

Wood Mechanical Behaviour

$$d_{\mathrm{m}} = 1 - \frac{\dfrac{\sigma_{3\mathrm{m}}}{k}\left(\delta_{\mathrm{m}} - \delta_{2\mathrm{m}}\right) + \delta_{1\mathrm{m}}\left(\delta_{3\mathrm{m}} - \delta_{\mathrm{m}}\right)}{\delta_{\mathrm{m}}\left(\delta_{3\mathrm{m}} - \delta_{2\mathrm{m}}\right)} \tag{2.54}$$

and in the last one (i.e. for $\delta_{3\mathrm{m}} \leq \delta_{\mathrm{m}} \leq \delta_{\mathrm{um}}$),

$$d_{\mathrm{m}} = 1 - \frac{\sigma_{3\mathrm{m}}\left(\delta_{\mathrm{um}} - \delta_{\mathrm{m}}\right)}{k\delta_{\mathrm{m}}\left(\delta_{\mathrm{um}} - \delta_{3\mathrm{m}}\right)} \tag{2.55}$$

Replacing the damage parameter into the constitutive equation (2.45) via diagonal matrix **D** permits the simulation of gradual damage propagation as a function of the current equivalent displacement δ_{m}.

As already referred, this model, or simpler versions of it, are widely used in the following sections dedicated to analysis of wood fracture characterisation under different mode loading.

References

Anderson, T. L. (2005). *Fracture Mechanics: Fundamentals and Applications*, Third Edition, CRC Press, Boca Raton, FL.

ASTM D3737-12 (2012). *Standard Practice for Establishing Allowable Properties for Structural Glued Laminated Timber (Glulam)*, ASTM International, West Conshohocken, PA.

de Moura, M. F. S. F., R. Fernandes, F. G. A. Silva and N. Dourado (2015). Mode II fracture characterization of a hybrid cork/carbon-epoxy laminate. *Compos Part B-Eng*, 76:44–51.

EN 408:2010+A1:2012. Timber structures—Structural timber and glued laminated timber— Determination of some physical and mechanical properties.

Gonçalves J. P. M., M. F. S. F. de Moura, P. M. S. T. de Castro and A. T. Marques (2000). Interface element including point-to-surface constraints for three-dimensional problems with damage propagation. *Eng Comput*, 17:28–47.

Hankinson, R. L. (1921). Investigation of crushing strength of spruce at varying angles of grain, Air Force Information Circular No. 259, U. S. Air Service.

Irwin, G. R. (1957). Analysis of stress and strains near the end of a crack traversing a plate. *ASME J Appl Mech*, 24:361–64.

Irwin, G. R. and J. A. Kies (1954). Critical energy rate analysis of fracture strength. *Weld J Res Suppl*, 33:193s.

Kolmann, F. P. F. and W. A. Côté (1984). *Principles of Wood Science and Technology: Solid Wood*, Springer-Verlag, Berlin.

Krueger, R. (2002). The virtual crack closure technique: History, approach and applications, NASA/ CR-2002-211628, Icase Report No. 2002–10.

Liu, J. Y. (1984). Evaluation of the tensor polynomial strenght theory of wood. *J Compos Mater*, 18:216–26.

Norris, C. B. (1962). Strength of orthotropic materials subjected to combined stresses. Report No. 1816, Forest Products Laboratory, Madison, WI.

Tsai, S. W. (1968). Strength theories of filamentary structures. In: *Fundamental Aspects of Fiber Reinforced Plastic Composites*, Conference Proceedings, R.T. Schwartz and H. S. Schwartz (Editors), Dayton, Ohio, 24–26 May 1966, Wiley Interscience, New York, 1968, pp. 3–11.

Tsai, S. W. and E. M. Wu (1971). A general theory of strength of anisotropic materials. *J Compos Mater*, 5:58–80.

Xavier J. C., N. M. Garrido, M. Oliveira, J. L. Morais, P. P. Camanho and F. Pierron (2004). A comparison between the Iosipescu and off-axis shear test methods for the characterization of *Pinus Pinaster* Ait. *Compos Part A-Appl Sci Manuf*, 35:827–40.

Xavier, J. C., M. Oliveira, J. L. Morais and T. Pinto (2009). Measurement of the shear properties of clear wood by the Arcan test. *Holzforschung*, 63:217–25.

Yoshihara, H., Y. Kubojima, K. Nagaoka and M. Ohta (1998). Measurement of the shear modulus of wood by static bending tests. *J Wood Sci*, 44:15–20.

3

Mode I Fracture Characterisation

As previously mentioned, on a macroscopic scale, wood is regarded as an orthotropic material exhibiting, in each point, three distinct directions of symmetry: the longitudinal (L) direction following the tracheids or fibre arrangement; the radial (R) direction regards the concentric growth rings and the tangential (T) direction (Figure 3.1). Hence, for a complete mode I fracture characterisation of wood, the critical energy release rates should be determined for the six independent fracture systems, i.e. LR, LT, RT, RL, TL and TR [with the first letter indicating the normal direction of the crack plane and the second specifying the direction of crack propagation (Figure 3.2)]. However, in the context of structural application of wood, the RL and TL fracture systems are the most important ones owing to propensity of crack propagation parallel to grain, with several experimental tests being proposed to estimate the critical energy release rate in mode I, i.e. G_{Ic}^{RL} and G_{Ic}^{TL}. Hence, among the most common fracture tests, one may point the double cantilever beam (DCB), the single-edge-notched beam loaded in three-point-bending (SEN-TPB), the tapered double cantilever beam (TDCB) and the compact tension test (CT) that will be analysed in the following sections.

3.1 Double Cantilever Beam

3.1.1 Test Description

Figure 3.3a shows the sketch of a DCB specimen with the initial crack length denoted as a_0, while Figure 3.3b presents the test setup employed in this work. The applied loading P at the specimen extremities leads to crack propagation under pure mode I (opening mode). Owing to its simplicity and ability to provide self-similar crack growth for a considerable extent, the DCB is the most appealing geometry to be used in wood fracture characterisation under pure mode I loading. Because wood is relatively weak when loaded perpendicularly to grain and fracture propagation in the LR and LT systems is retained by the fibres disposal, it turns that the RL and TL fracture systems are the most studied ones.

Because of characteristic damage mechanisms, such as micro-cracking and fibre-bridging, the monitoring of crack length in wood is difficult to accomplish with the necessary accuracy (de Moura et al., 2006). To overcome this limitation, an equivalent data reduction scheme based on the beam theory and crack equivalent concept was developed to evaluate fracture toughness in wood.

3.1.2 Classical Data Reduction Schemes

The evaluation of the fracture energy is accomplished using the load–displacement (P–δ) curve ensued from the DCB test and the Irwin–Kies expression (Eq. 2.39). To get dC/da, it

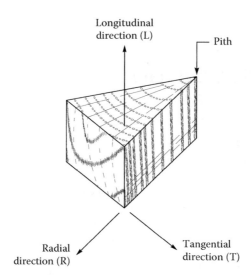

FIGURE 3.1
Wood anatomical directions: longitudinal, radial and tangential (Dourado et al., 2015).

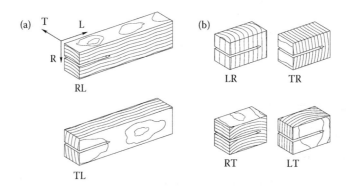

FIGURE 3.2
Fracture systems in wood. (Adapted from Dourado et al., 2010.)

is necessary to establish the $C = f(a)$ relationship, which requires the crack-length monitoring. This is known as the compliance calibration method (CCM). Alternatively, the corrected beam theory (Timoshenko) may also be used (ISO 15024:2001),

$$G_I = \frac{3P\delta}{2B(a+|\Delta|)} \quad (3.1)$$

where Δ is an experimentally determined correction term that envisages to account for root rotation effects at the crack tip. Thus, it becomes evident that the measurement of the crack length a in classical data reduction schemes is a fundamental issue regarding the identification of fracture energy.

As already referred, the crack length monitoring in wood is not easy to perform with necessary accuracy (Dourado et al., 2010) (Figure 3.4). In fact, toughening mechanisms like micro-cracking, crack-branching or fibre bridging, which take place in the vicinity of the crack tip (de Moura et al., 2008, Dourado et al., 2008) give rise to the development of

Mode I Fracture Characterisation

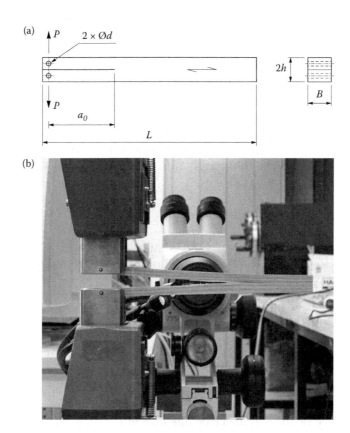

FIGURE 3.3
(a) Sketch of the double cantilever beam (DCB) and (b) the DCB fracture test.

FIGURE 3.4
Crack-tip detail in the RL fracture system (Dourado et al., 2010).

a non-negligible fracture process zone (FPZ). As a consequence, classical data reduction schemes based on crack length measurements can lead to important errors on the evaluated values of G_{Ic}. To overcome this limitation, two methods based on specimen compliance and the crack equivalent concept are presented in the following section.

3.1.3 Modified Experimental Compliance Method (MECM)

This method relies on the empirical relation for the compliance,

$$C = ka^n \qquad (3.2)$$

with k and n parameters (i.e. constants) to be evaluated experimentally. Hence, combining Eqs. (3.2) and (2.39) and bearing in mind that $C = \delta/P$ yields

$$G_{\mathrm{I}} = \frac{nP\delta}{2Ba} \qquad (3.3)$$

This data reduction scheme can be modified to evaluate the *Resistance*-curve (*R*-curve) from the experimental P–δ curve without direct measurement of crack length during crack propagation. However, the experimental compliance calibration is needed, which requires the execution of experimental tests with different initial crack lengths a_0. Therefore, Eq. (3.2) can be rewritten as

$$\log_{10} C_0 = n\log_{10} a_0 + \log_{10} k \qquad (3.4)$$

As a result, the constants n and k can be obtained on the linear regression basis and used in the subsequent fracture test. Thus, applying Eq. (3.4) for the current compliance C, the equivalent crack length a_{e} is estimated for each point of the P–δ curve (employing a_{e} and the actual compliance C values instead of a_0 and C_0). Then, the energy release rate $G_{\mathrm{I}}(a_{\mathrm{e}})$ is estimated by means of Eq. (3.3) (using a_{e} instead of a). Using this method, a complete *R*-curve is obtained.

3.1.4 Compliance-Based Beam Method (CBBM)

This method uses the Timoshenko beam theory to establish the $C = f(a)$ relationship. The specimen arms are considered as two cantilever beams clamped at the crack tip. The elastic strain energy of the cantilever beams due to bending including shear effects is defined as

$$U = 2\left[\int_0^a \frac{M_{\mathrm{f}}^2}{2E_{\mathrm{L}}I}\,dx + \int_0^a \int_{-h/2}^{h/2} \frac{\tau^2}{2G_{\mathrm{LR}}}\,Bdy\,dx \right] \qquad (3.5)$$

where M_{f} is the bending moment, I the second moment of the cross-section area, E_{L} the longitudinal elastic modulus and G_{LR} the shear modulus in the RL plane (for the case of the RL fracture system). The shear stress is obtained by

$$\tau = \frac{3}{2}\frac{V}{Bh}\left(1 - \frac{y^2}{c^2}\right) \qquad (3.6)$$

being c and V the beam half thickness and the transverse load that acts in each beam ($0 \le x \le a$). Then, the resulting displacement δ is determined applying the Castigliano theorem,

$$\delta = \frac{\partial U}{\partial P} = \frac{8Pa^3}{E_{\mathrm{L}}Bh^3} + \frac{12Pa}{5BhG_{\mathrm{LR}}} \qquad (3.7)$$

which can also be established in terms of compliance,

$$C = \frac{8a^3}{E_{\mathrm{L}}Bh^3} + \frac{12a}{5BhG_{\mathrm{LR}}} \qquad (3.8)$$

Mode I Fracture Characterisation

It should be noted that Eq. (3.8) does not take into account the root rotation effects and stress concentrations at the crack tip that occur in reality. Therefore, to account for these issues a correction Δ of the initial crack length has to be performed considering three different initial crack lengths (being $a_{0_1} > a_{0_2} > a_{0_3}$) in the same specimen (Figure 3.5). This means that for the formerly two larger pre-cracks (i.e. a_{0_1} and a_{0_2}) very slight loads should be applied to prevent damage onset, since the goal is only to measure the respective initial compliance values. Consequently, the fracture test is performed for the shortest pre-crack length (i.e. a_{0_3}). Noticeably, the pair of holes referred to as B and C in Figure 3.5 (i.e. for crack length equal to a_{0_2} and a_{0_3}, respectively) have to be performed after the compliance measurement made for previous longer pre-crack lengths. This procedure leads to the attainment of parameter Δ by means of a linear regression accomplished on experimental $C_0^{1/3}$ versus a_0 data points, as shown in Figure 3.6.

It is known that wood presents remarkable scatter in its elastic properties. To account for this disadvantage, a corrected flexural modulus E_f can be used instead of E_L in Eq. (3.8). Hence, the determination of E_f is carried out considering the corrected pre-crack $(a_0 + \Delta)$,

$$E_f = \left(C_0 - \frac{12(a_0 + \Delta)}{5BhG_{LR}} \right)^{-1} \frac{8(a_0 + \Delta)^3}{Bh^3} \tag{3.9}$$

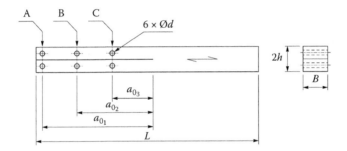

FIGURE 3.5
DCB used in the CBBM (Dourado et al., 2010).

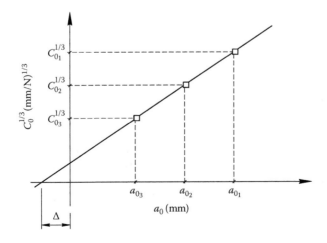

FIGURE 3.6
Method used to estimate the parameter Δ (Dourado et al., 2010).

On the other hand, with the crack advance, the relation between the equivalent crack length a_e and the current compliance (i.e. $C = \delta/P$) is obtained from Eq. (3.8) using E_f instead of E_L. Following this procedure, the effect of the FPZ is taken into account, since it is known that its presence affects the current compliance. The solution for the resulting cubic equation may be determined considering the general form of Eq. (3.8),

$$\alpha a_e^3 + \beta a_e + \gamma = 0 \tag{3.10}$$

being the coefficients,

$$\alpha = \frac{8}{Bh^3 E_f} ; \ \beta = \frac{12}{5BhG_{LR}} ; \ \gamma = -C \tag{3.11}$$

Then, using the Matlab® software, and keeping the real solution, yields

$$a_e = \frac{1}{6\alpha} A - \frac{2\beta}{A} \tag{3.12}$$

with,

$$A = \left[\left(-108\gamma + 12\sqrt{3\left(\frac{4\beta^3 + 27\gamma^2\alpha}{\alpha} \right)} \right) \alpha^2 \right]^{\frac{1}{3}} \tag{3.13}$$

The combination of Eqs. (2.39) and (3.8) leads to

$$G_I = \frac{6P^2}{B^2 h} \left(\frac{2a_e^2}{E_f h^2} + \frac{1}{5G_{LR}} \right) \tag{3.14}$$

Therefore, the measurement of the current crack length a becomes unnecessary and the elastic properties variability among different specimens is accounted for. The presented method also provides a complete R-curve.

3.1.5 Numerical Validation

A common way to perform the validation of data reduction schemes consists in verifying whether the plateau value of the R-curve ensued from a given method [e.g. modified experimental compliance method (MECM) or compliance-based beam method (CBBM)] agrees with the critical energy release rate (i.e. G_{Ic}inp) used as input in cohesive zone modelling (CZM) (Section 2.3.2). Hence, the simulation with CZM of the DCB test was performed (Figure 3.7) considering (Figure 3.5) $L = 240$, $B = 20$, $h = 10$ and $a_{0_1} = 160$, $a_{0_2} = 130$, $a_{0_3} = 100$ (all dimensions in mm). Axes x and y shown in Figure 3.7 are coincident with wood anatomic directions L and R, respectively (Figure 3.2), mimicking the definition of the RL fracture propagation system. The damage law used in the CZM was the bilinear one (Figure 3.8), which results from the trapezoidal relation shown in Figure 2.16, considering $\delta_{2,I} = \delta_{1,I}$. de Moura et al. (2008) have identified the following damage parameters for *Pinus pinaster* Ait.: $\sigma_{1,I} = 5.34$ MPa, $\delta_{3,I} = 0.076$ mm and $\sigma_{3,I} = 0.49$ MPa using the DCB test.

Mode I Fracture Characterisation

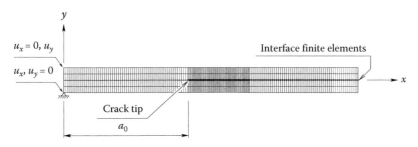

FIGURE 3.7
Finite element mesh of the DCB test. (Adapted from Dourado et al., 2010.)

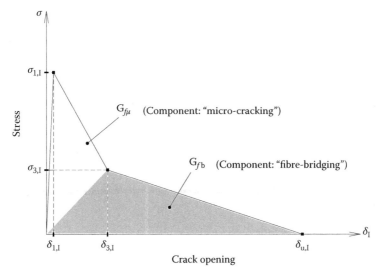

FIGURE 3.8
Bilinear softening cohesive law ($G_{Ic} = G_{f\mu} + G_{fb}$). (Adapted from de Moura et al., 2008.)

The set of mechanical properties (elastic and fracture parameters) used in the numerical simulation is presented in Table 3.1.

Figure 3.9 shows the $P-\delta$ curve obtained in the finite element analysis (FEA) of the DCB test, considering the initial crack length ($a_0 = 100$ mm), together with the numerical response using initial crack lengths a_0 in the interval of 110–160 mm. Following MECM, this data was then used to get the embedded numerical compliance calibration curve on the $\log_{10} C_0 - \log_{10} a_0$ basis according to Eq. (3.4). The obtained law was subsequently used to assess the equivalent crack length a_e associated to each point of the numerical $P-\delta$ curve shown in Figure 3.9, using Eq. (3.2) as a function of the current compliance ($C = \delta/P$). Then, the R-curve (i.e. $G_I(a_e)$) was evaluated employing Eq. (3.3) using a_e instead of a (Figure 3.10).

TABLE 3.1
Mechanical Properties Used in the FEA for *P. pinaster* Ait

E_L (GPa)	E_R (GPa)	ν_{LR}	G_{LR} (GPa)	$\sigma_{1,I}$ (MPa)	$\sigma_{1,II}$ (MPa)	G_{Ic} (N/mm)	G_{IIc} (N/mm)
12.0	1.91	0.47	1.12	7.9	16.0	0.264	0.63

Source: Oliveira et al. (2007).

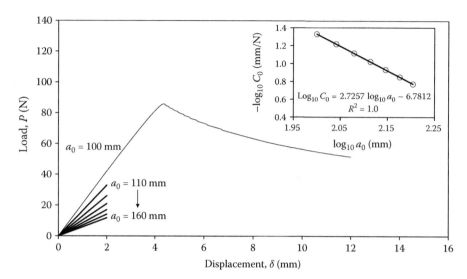

FIGURE 3.9
Load–displacement curves obtained in the FEA of the DCB and the resulting compliance calibration numerical curve ($n = 2.7257$). (Adapted from Dourado et al., 2010.)

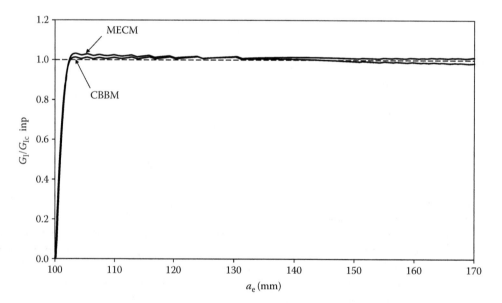

FIGURE 3.10
R-curves estimated by numerical simulation of the DCB (Dourado et al., 2010).

The presented $P-\delta$ curve of the DCB (Figure 3.9) was furthermore treated according to the CBBM, using the correction factor Δ determined as illustrated in Figure 3.6. Then, computing the flexural modulus E_f as a function of the initial conditions (i.e. a_0 and C_0), a corresponding R-curve was obtained (Eq. 3.14).

For the sake of simplicity, the R-curves ensued from a validation procedure are commonly normalised by the critical energy release rate used as input (i.e. $G_{Ic\ inp} = 0.264$ N/mm in

Mode I Fracture Characterisation

Table 3.1). By doing so for the R-curves obtained for the MECM and CBBM (i.e. $G_I/G_{Ic\ inp}(a_e)$), it becomes clear that the CBBM captures with excellent accuracy the value of the fracture energy used as input (Figure 3.10). On the other hand, the MECM overestimates the input fracture energy for some extent at the beginning of the plateau, though the value used as input is well captured after some propagation extent.

Summing up, it can be concluded that both data reduction schemes (i.e. MECM and CBBM) are appropriate when applied to the DCB test in wood.

3.1.6 Experimental and Numerical Results

Figure 3.11 shows a set of experimental $P-\delta$ curves obtained under displacement control (5 mm/min), using the following nominal dimensions (Figure 3.3a): $L = 280$, $2h = B = 20$ and $a_0 = 100$ (mm). These curves put into evidence a non-negligible scatter that exists in wood, which results mainly from the elastic modulus (i.e. the longitudinal Young modulus, E_L). Consequently, the evaluation of the R-curves might take into account this variability by considering the elastic response of each specimen. As observed in Section 3.1.4, the CBBM is adequate to deal with this specificity, since the formulation requires the determination of the flexural modulus (E_f) of each specimen as a function of the initial conditions (i.e. a_0 and C_0). Figure 3.12 shows the plotting of the resulting R-curves obtained by the CBBM. Table 3.2 presents the resume of the main results ensued from the DCB tests, i.e. the ultimate load (P_{max}) and the critical energy release rate (G_{Ic}), using the equivalent data reduction schemes presented in Sections 3.1.3 and 3.1.4. It should be observed that the registered scatter attained for G_{Ic} can be considered quite acceptable for biological materials (usually not far from the interval of 20%).

Finally, the material cohesive properties (i.e. $\delta_{3,I}$, $\sigma_{3,I}$, $\sigma_{1,I}$ and G_{Ic} according to Figure 3.8) were identified through an inverse method that seeks the agreement between the numerical and experimental $P-\delta$ curves (see Figure 3.13), by means of a genetic algorithm (de Moura et al., 2008). The ensued numerical $P-\delta$ curve was then used to evaluate the corresponding R-curve to verify whether the value of G_{Ic} used as input in the cohesive model (i.e. $G_{Ic\ inp}$) could accurately reproduce the numerical results that had

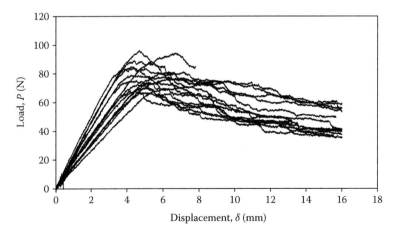

FIGURE 3.11
Experimental $P-\delta$ curves obtained in the DCB test (de Moura et al., 2008).

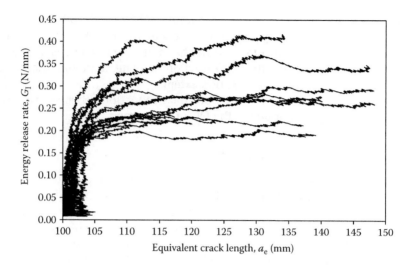

FIGURE 3.12
Experimental R-curves obtained in the DCB test using the CBBM.

TABLE 3.2

Resume of Results Obtained Experimentally with the DCB Tests

	Experimental Results		Numerical Results		Error	
Specimen	P_{max} (N)	G_{Ic} (N/mm)	P_{max} (N)	G_{Ic} (N/mm)	P_{max} (%)	G_{Ic} (%)
1	83.8	0.250	81.8	0.247	−2.43	−1.2
2	68.5	0.180	67.4	0.178	−1.61	−1.1
3	94.4	0.350	91.0	0.343	−3.62	−2.0
4	79.0	0.254	78.9	0.251	−0.08	−1.2
5	75.3	0.210	74.9	0.207	−0.52	−1.4
6	89.2	0.290	90.1	0.282	1.01	−2.8
7	77.7	0.365	77.9	0.358	0.21	−1.9
8	80.5	0.320	75.9	0.315	−5.73	−1.6
9	74.2	0.270	74.0	0.265	−0.32	−1.9
10	72.6	0.210	72.0	0.208	−0.73	−1.1
11	66.8	0.230	67.3	0.227	0.68	−1.3
12	73.0	0.350	71.7	0.348	−1.73	−0.7
13	84.8	0.220	84.7	0.218	−0.15	−1.0
14	96.2	0.265	94.4	0.265	−1.89	0.0
15	70.3	0.200	67.9	0.199	−3.50	−0.8
Average values	79.1	0.264	78	0.261		
St. deviation	9.1	0.059	8.8	0.058		

Source: de Moura et al. (2008).

been achieved with the CBBM. Figure 3.14 allows concluding that the entire methodology is robust, since the value of $G_{Ic\,inp}$ is well captured by the horizontal asymptote of the R-curve.

As a final remark, it can be concluded that the DCB test with the CBBM is a valuable methodology to perform mode I fracture characterisation of wood, when propagation takes place in the direction parallel to grain (i.e. RL and TL fracture systems).

Mode I Fracture Characterisation

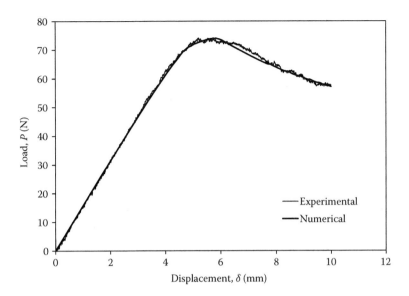

FIGURE 3.13
Agreement between numerical and experimental $P-\delta$ curves. (Adapted from de Moura et al., 2008.)

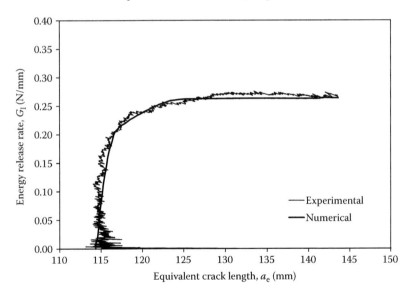

FIGURE 3.14
Agreement between numerical and experimental R-curves. (Adapted from de Moura et al., 2008.)

3.2 Single-Edge-Notched Beam Loaded in Three-Point-Bending

3.2.1 Test Description

The single-edge-notched beam loaded in three-point-bending (Figure 3.15) is a convenient and easy test to execute, ideally allowing performing fracture tests in any wood fracture system (Figure 3.2). This specimen, originally proposed by Gustafsson (1988), for wood constitutes a composite beam formed by three segments bonded to each other by means of

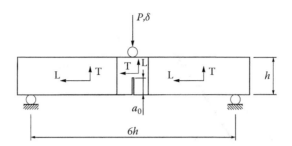

FIGURE 3.15
SEN-TPB setup.

a suitable structural adhesive. Hence, two longitudinal segments of equal size (Figure 3.16) are stuck to a central part conveniently oriented as to assure crack propagation in a pretended fracture system (Figure 3.2). This characteristic constitutes an advantage of this specimen configuration when compared with the DCB, whose application is restricted to propagations along the grain direction. Figure 3.17 shows a set of load–displacement curves obtained in the SEN-TPB test following the relations presented in Figure 3.16a

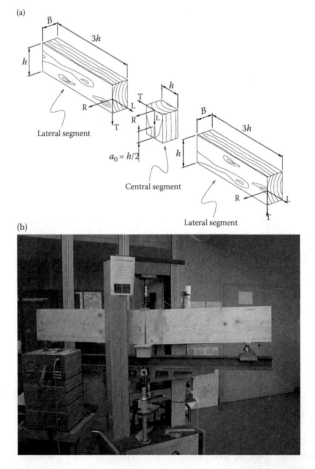

FIGURE 3.16
(a) Layout of the SEN-TPB for the TL fracture system and (b) SEN-TPB fracture test.

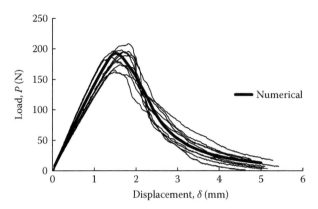

FIGURE 3.17
Experimental P–δ curves obtained in the SEN-TPB test for *Picea abies* L. (Dourado et al., 2015).

(i.e. initial crack/depth ratio a_0/h set to 0.5 and span/depth ratio to 6) for $h = 140$ mm. In the present case, the TL fracture system was considered instead of the RL, owing to the easiness to get larger specimen sizes. The selected wood was *Picea abies* L., taking advantage of the fact that notches in this wood species are more regularly distributed along the grain direction and exhibit smaller sizes.

3.2.2 Data Reduction Scheme Based on Equivalent LEFM

In the framework of the equivalent linear elastic fracture mechanics (eqLEFM), the (equivalent) crack length in a linear elastic solid domain is assumed as the one that provides the same compliance as that revealed by the solid with a damaged zone (Morel et al., 2005). This is performed by first assessing the compliance evolution of such solid as a function of the crack length (i.e. $C(a)$). This procedure may be conducted numerically (Figure 3.18) through linear elastic finite element analyses (FEA) using characteristic (reference) material elastic properties of wood (Figure 3.19) and several values of a, leading to a polynomial function $C(a)$, as in the CCM. Typical scatter in wood elastic properties is responsible for the eventual dissimilarity observed in the numerical and experimental compliances. However, it is possible to perform a correction on these values by means of a multiplicative correction factor estimated using the initial conditions as follows: $\psi = C_{\exp}(a_0)/C(a_0)$, with

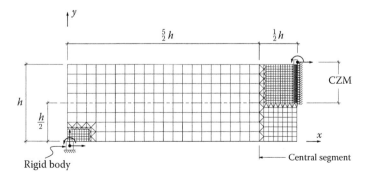

FIGURE 3.18
Mesh used in the numerical simulations of the SEN-TPB. (Adapted from de Moura et al. 2010.)

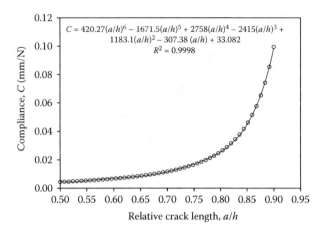

FIGURE 3.19
Compliance calibration function of the SEN-TPB (Figure 3.18).

$C(a_0)$ being the numerical compliance obtained for the initial crack length a_0. Hence, the evaluation of the equivalent crack length (a) is accomplished by identifying the value of a that allows matching the corrected value of the compliance $C_{cor}(a) = \psi C(a)$ to the experimental one $(C_{exp}(a))$. This procedure may be conducted by means of numerical methods (e.g. the bisection method). It has been observed by Morel et al. (2005) that the compliance of the SEN-TPB is mainly dictated by the elastic modulus of the central segment (i.e. E_T in Figure 3.16a). Therefore, in the FEA, it suffices to modify the value of E_T to attain the experimental compliance.

The strain energy release rate $G_I(a)$ in a solid of width B is evaluated dividing the elastic strain energy $w(a)$ released by the propagation of an infinitesimal crack extent δa by the corresponding area,

$$G_I(a) = \frac{w(a)}{B\delta a} \tag{3.15}$$

The elastic strain energy $w(a)$ represented by the dashed area in Figure 3.20 is computed numerically from the load–displacement curve using two straight lines that correspond to equivalent crack lengths $a - \delta a/2$ and $a + \delta a/2$. These two lines denote corrected values of compliance [i.e. $C_{cor}(a - \delta a/2)$ and $C_{cor}(a + \delta a/2)$] obtained following the aforementioned

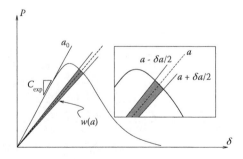

FIGURE 3.20
Evaluation of the elastic energy release rate $w(a)$. (Adapted from Morel et al., 2005.)

Mode I Fracture Characterisation

procedure. According to the fundamentals of LEFM, the equivalent crack extent δa that should be used to evaluate $G_I(a)$ has to be small [i.e. 1% of the initial crack length as observed by Morel et al. (2002, 2003)]. This method (designated by Coureau et al., 2013 as *eqLEFM*) renders possible evaluating the R-curve (Figure 3.21) without performing the derivative of the compliance function (Eq. 2.39), which is often at the source of non-negligible errors.

The validation of the presented data reduction scheme was performed numerically considering CZM. Hence, for $h = 140$ mm in Figure 3.18, the cohesive parameters representative of *Picea abies* L. (i.e. $\delta_{3,I} = 0.085$ mm, $\sigma_{3,I} = 0.3$ MPa, $\sigma_{1,I} = 2.0$ MPa and $G_{Ic} = 0.152$ N/mm as represented in Figure 3.8) are identified by Dourado et al. (2015), and for the elastic properties presented in Table 3.3, a load–displacement curve is determined (Figure 3.17). Then, using the procedure detailed earlier, the corresponding R-curve is obtained (Figure 3.21). Since a good agreement between the numerical and experimental R-curves has been achieved (Figure 3.21), the method is considered valid to perform the evaluation of fracture toughness in wood.

3.2.3 Compliance-Based Beam Method

The existence of a material discontinuity in the central section of the specimen (Figure 3.15) affects the stress profile in that region, impeding the direct application of beam theory to the SEN-TPB. However, this limitation is overwhelmed following the suggestion of Kienzler and Herrmann (1986), who considered the existence of a stress relief region (SRR) of a triangular shape localised near the existing crack (Figure 3.22). In the course of the loading process, stresses in this region are dropped to zero, which is considered in the assessment of the strain energy release due to bending,

$$U = 2\left[\int_0^{L_2} \frac{M_f^2}{2E_L I} dx + \int_{L_2}^{L_1} \frac{M_f^2}{2E_L I_{sr}} dx + \int_{L_1}^{L} \frac{M_f^2}{2E_T I_{sr}} dx\right] \tag{3.16}$$

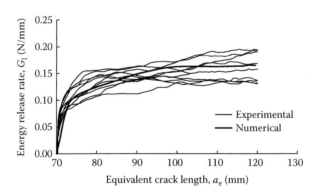

FIGURE 3.21
Numerical agreement of R-curves obtained with the SEN-TPB test using *eqLEFM*.

TABLE 3.3
Mechanical Properties Used in the FEA for *Picea abies* L

E_L (GPa)	E_R (GPa)	E_T (GPa)	ν_{TL}	ν_{RL}	ν_{TR}	G_{TL} (GPa)	G_{RT} (GPa)	G_{RL} (GPa)
9.90	0.73	0.41	0.018	0.032	0.306	0.610	0.022	0.500

Source: Dourado et al. (2008).

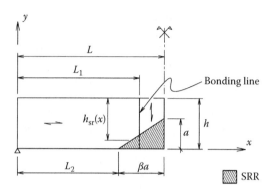

FIGURE 3.22
Simplified sketch of the SEN-TPB showing the stress relief region (SRR). (Adapted from Dourado et al., 2015.)

According to Figure 3.22, L represents the half-length of the central section, L_1 is the span of the lateral segment, M_f stands for the bending moment ($M_f = Px/2$) and E_L and E_T for the moduli of elasticity in the longitudinal and transverse direction, respectively. Furthermore, the quantity L_2 follows on the length of a cathetus of the SRR, as $L_2 = L - \beta a$, with β standing for a multiplicative factor used to define the size of the SRR (to be discussed later). Parameters I and I_{sr} are the second moment of area of the entire cross section (height h) and of the section within the SRR $\left(I = Bh^3/12 \text{ and } I_{sr} = B[h_{sr}(x)]^3/12\right)$, respectively. Also, B is the specimen width (Figure 3.16a) and

$$h_{sr}(x) = h + \frac{a}{L - L_2}(L_2 - x) \tag{3.17}$$

with a being the crack length. Hence, combining the earlier referred expressions of the second moment of area and Eq. (3.17) with Eq. (3.16), and applying the Castigliano theorem (i.e. $\delta = \partial U/\partial P$), the specimen compliance ($C = \delta/P$) becomes

$$C = \frac{2L_2^3}{E_L B h^3} + \frac{6(L - L_2)}{Ba} \left\{ \frac{1}{E_L} \left[\frac{L_1^2}{2(h_{sr}(L_1))^2} - \frac{L_2^2}{2h^2} + \frac{L - L_2}{a} \left(\frac{L_2}{h} - \frac{L_1}{h_{sr}(L_1)} \right) \right. \right.$$

$$\left. + \left(\frac{L - L_2}{a} \right)^2 \ln \frac{h}{h_{sr}(L_1)} \right] + \frac{1}{E_T} \left[\frac{L^2}{2(h - a)^2} - \frac{L_1^2}{2(h_{sr}(L_1))^2} \right.$$

$$\left. \left. + \frac{L - L_2}{a} \left(\frac{L_1}{h_{sr}(L_1)} - \frac{L}{h - a} \right) + \left(\frac{L - L_2}{a} \right)^2 \ln \frac{h_{sr}(L_1)}{h - a} \right] \right\} \tag{3.18}$$

As already discussed, wood is often affected by scatter in its elastic properties. Due to the arrangement of the constitutive wood parts shown in Figure 3.15, specimen compliance is more sensible to the modulus of elasticity in the transverse direction E_T than to the longitudinal one (i.e. E_L). As a consequence, the referred inaccuracy can be corrected by considering an equivalent quantity known by flexural modulus (E_{Tf}), as an alternative to the nominal value of E_T. This quantity is determined using Eq. (3.18) by accounting for the initial conditions, i.e. a_0 and C_0 instead of a and C, respectively.

Mode I Fracture Characterisation

Since crack length monitoring in wood is very difficult to perform with accuracy, an equivalent value (a_e) is considered instead, defined as a function of the current compliance (Eq. 3.18). This method follows the so-called *eqLEFM*, which is used to deal with quasi-brittle failure. This means that the compliance evolution is due to the totality of non-linear phenomena associated to the FPZ development and/or crack propagation.

For the reason that Eq. (3.18) does not provide the analytical solution for the equivalent crack length (a_e) as a function of the compliance C, the problem has to be solved numerically (e.g. bisection method). Hence, using the Irwin–Kies equation (Eq. 2.39) and Eq. (3.18), the *R*-curve $(G_I = f(a_e))$, can be obtained from

$$
\begin{aligned}
G_I = {} & \frac{3P^2(L-L_2)}{B^2}\Bigg\{\frac{1}{E_L}\Bigg[\frac{L_2^2}{2h^2a_e^2} - \frac{L_1^2(h+3\eta a_e)}{2a_e^2(h+\eta a_e)^3} + \frac{(L-L_2)}{a_e^3}\Bigg(\frac{L_1(2h+3\eta a_e)}{(h+\eta a_e)^2} - \frac{2L_2}{h}\Bigg)\\
& - \frac{(L-L_2)^2}{a_e^3}\Bigg(\frac{3}{a_e}\ln\frac{h}{h+\eta a_e} + \frac{\eta}{h+\eta a_e}\Bigg)\Bigg] + \frac{1}{E_{Tf}}\Bigg[\frac{L^2(3a_e-h)}{2a_e^2(h-a_e)^3} + \frac{L_1^2(h+3\eta a_e)}{2a_e^2(h+\eta a_e)^3}\\
& - \frac{(L-L_2)}{a_e^3}\Bigg(\frac{L_1(2h+3\eta a_e)}{(h+\eta a_e)^2} + \frac{L(3a_e-2h)}{(h-a_e)^2}\Bigg)\\
& + \frac{(L-L_2)^2}{a_e^3}\Bigg(\frac{h(1+\eta)}{(h-a_e)(h+\eta a_e)} - \frac{3}{a_e}\ln\frac{h+\eta a_e}{h-a_e}\Bigg)\Bigg]\Bigg\}
\end{aligned}
\tag{3.19}
$$

with $\eta = (L_2 - L_1)/(L - L_2)$. The evaluation of G_I using Eq. (3.19) requires the continuous identification of L_2 which depends on β, i.e. $L_2 = L - \beta a$ (Figure 3.22). To define appropriate values of β, a numerical analysis was performed by de Moura et al. (2010) and is described in the following section.

3.2.4 Numerical Validation of the Compliance-Based Beam Method

A numerical validation of the proposed data reduction method has been accomplished by comparing the horizontal asymptote of the *R*-curve ensued from the CBBM presented in Section 3.2.3, with the value of the critical energy release rate that has been used as input (i.e. $G_{Ic\,inp}$) in the CZM. This procedure was preceded by the evaluation of the multiplicative factor β used to estimate the length of the horizontal cathetus of the SRR represented in Figure 3.22 (i.e. $\beta = (L - L_2)/a_e$), as a function of the equivalent crack length a_e. Hence, six geometrically similar specimens with different sizes (Table 3.4) were modelled

TABLE 3.4

Dimensions of the SEN-TPB Specimens according to Figure 3.22

Series	h (mm)	L_1 (mm)	L (mm)	a_0 (mm)	B (mm)
h_1	17.5	43.75	52.5	8.75	5
h_2	35	87.5	105	17.5	10
h_3	70	175	210	35	20
h_4	140	350	420	70	40
h_5	280	700	840	140	80
h_6	560	1400	1680	280	160

(Figure 3.18), for the material disposal represented in Figures 3.15 and 3.16 and elastic properties listed in Table 3.3. The numerical modelling was executed using cohesive parameters representative of *Picea abies* L. (i.e. $\delta_{3,I} = 0.09$ mm, $\sigma_{3,I} = 0.03$ MPa, $\sigma_{1,I} = 1.66$ MPa and $G_{Ic} = 0.144$ N/mm, according to Figure 3.8) that have been identified by Dourado et al. (2008). Figure 3.23 shows the set of *R*-curves obtained with this method normalised by the critical energy release rate used as input, $G_{Ic\,inp}$. The value of β was identified as the one that provides a clear plateau on the *R*-curve for each specimen size. It is possible to conclude that the multiplicative factor β has converged to a value of 1.07 for specimen sizes (i.e. h) higher or equal than 70 mm. This value agrees with the conclusions of Kienzler and Herrmann (1986), who stated that the SRR is well described by an isosceles triangle. Conversely, specimen sizes smaller than 70 mm are inadequate to perform fracture toughness evaluations in the tested wood species (i.e. *Picea abies* L.), since the corresponding *R*-curves do not attain a horizontal asymptote for any value of β. A way to testify

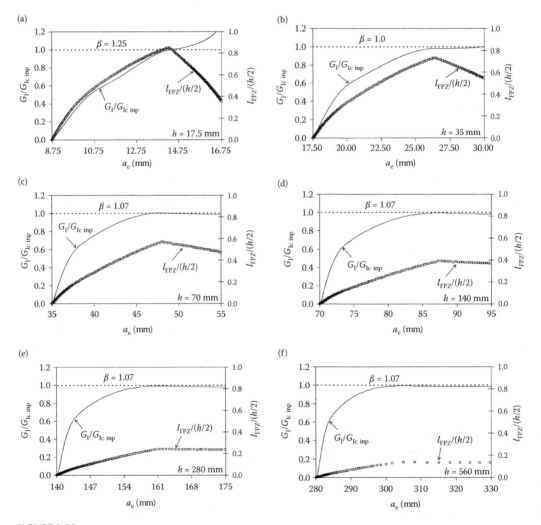

FIGURE 3.23
R-curves for the six modelled specimen sizes presented in Table 3.4: (a) h_1, (b) h_2, (c) h_3, (d) h_4, (e) h_5 and (f) h_6 (Dourado et al., 2010).

Mode I Fracture Characterisation

this limitation more accurately consists in observing the trend of the developed cohesive zone length (CZL) (l_{FPZ}) throughout loading, as represented in Figure 3.23 (normalised by the initial ligament length $h/2$). It is clear that specimens smaller than 140 mm height (Figure 3.23a–c) do not allow the attainment of the so-called self-similar crack growth, being inappropriate for fracture characterisation of this material with this test. In fact, in these smaller specimens, the ligament length is insufficient to provide self-similar crack growth due to development of normal compressive stresses induced by bending above the specimen neutral axis. These compressive stresses interact with the FPZ development, leading to a spurious increase of fracture energy.

As a final remark, it can be stated that this test is appropriate for wood fracture characterisation under mode I loading, although special attention might be paid to specimen size, in particular its ligament length.

3.2.5 Experimental and Numerical Results

Figure 3.24(a)–(c) reveals the numerical agreement obtained for the R-curves of series h_3, h_4 and h_5 (Table 3.4) using the CBBM. With the exception of the specimen size of series h_3 (i.e. $h = 70$ mm), clear plateaus were identified, leading to the accurate evaluation of the energy release rate G_{Ic} in wood (*Picea abies* L.). Figure 3.24 also shows the energy release rate obtained for the ultimate load [$G_I(P_u)$], emphasising that in the post-peak regime the energy release rate in wood continues to increase beyond P_u. Tables 3.5 and 3.6 present the resume of the main experimental results obtained for series h_4 and h_5. The normalisation of G_{Ic} and $G_I(P_u)$ by the basic density ρ of the specimen aims disregarding the contribution of this property to the fracture energy release rate.

The excellent agreement between the numerical and experimental R-curves shown in Figure 3.24 reveals the reliability of the proposed procedure when applied to SEN-TPB test for fracture characterisation of wood under mode I loading.

3.3 Tapered Double Cantilever Beam

3.3.1 Test Description

This specimen was extensively used by Morel and co-authors (1998, 2000, 2002, 2003) to characterise wood (*Picea abies* L. and *Pinus pinaster* Ait.) mode I fracture, as it provides stable crack growth through a constant crack speed propagation. The specimen geometry (Figure 3.25) is defined to match the longitudinal orientation with the grain direction of wood, with a slender increase on the cross section along the crack extent, i.e. $h = f(x)$. The goal is to design the specimen so that the compliance changes linearly with crack length leading to a constant dC/da relation (Davalos et al., 1998) for a given crack extension. The consequence is that strain energy release rate G becomes independent of crack length a, i.e. it varies only as a function of the applied load during crack growth, which simplifies dramatically the fracture toughness evaluation (Mostovoy et al., 1967).

3.3.2 Data Reduction Scheme

Mostovoy et al. (1967) were the first to propose the use of a so-called height tapered DCB (TDCB) test configuration to measure the resistance to crack growth in adhesive joints considering isotropic materials. The compliance of the beam was determined, including the

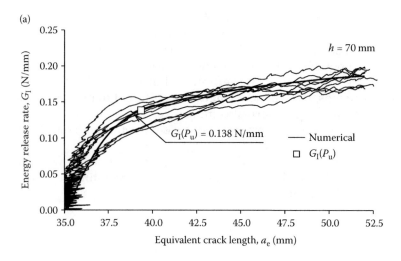
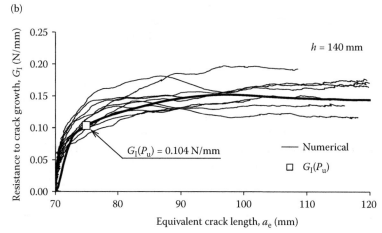
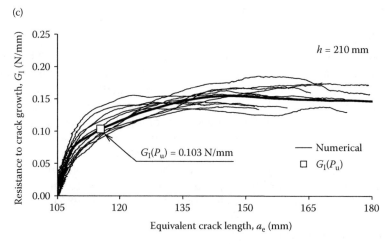

FIGURE 3.24
Experimental and numerical *R*-curves for series: (a) h_3, (b) h_4 and (c) h_5 (in *Picea abies* L.) (Dourado et al., 2015).

Mode I Fracture Characterisation

TABLE 3.5

Resume of Experimental Results Ensued from Series h_4 (i.e. $h = 140\,\text{mm}$)

Specimen	C_0 (×10⁻³ mm/N)	P_u (N)	$G_I(P_u)$ (N/mm)	$G_I(P_u)/\rho$ (m³/s²)	G_{Ic} (N/mm)	G_{Ic}/ρ (m³/s²)
1	7.7	189.1	0.12	0.32	0.13	0.32
2	7.1	207.8	0.13	0.37	0.16	0.44
3	6.4	227.3	0.14	0.35	0.15	0.37
4	6.3	185.7	0.09	0.25	0.14	0.38
5	7.8	194.8	0.12	0.33	0.14	0.33
6	6.2	198.1	0.12	0.33	0.17	0.46
7	8.2	173.8	0.10	0.30	0.12	0.30
8	5.9	244.2	0.15	0.35	0.17	0.40
9	7.6	180.1	0.09	0.19	0.15	0.32
10	4.9	237.8	0.12	0.24	0.19	0.39
Average	6.8	203.9	0.12	0.30	0.15	0.37
CoV (%)	15.3	12.1	17.5	19.2	14.0	14.5

C_0: initial compliance; P_u: ultimate load; G_{Ic}: critical energy release rate; $G_I(P_u)$: resistance to crack growth at the ultimate load; ρ: wood basic density (in *Picea abies* L.).
Source: Dourado et al. (2015).

TABLE 3.6

Resume of Experimental Results Ensued from Series h_5 (i.e., $h = 210\,\text{mm}$)

Specimen	C_0 (×10⁻³ mm/N)	P_u (N)	$G_I(P_u)$ (N/mm)	$G_I(P_u)/\rho$ (m³/s²)	G_{Ic} (N/mm)	G_{Ic}/ρ (m³/s²)
1	4.6	321.2	0.1	0.27	0.16	0.42
2	4.9	365.3	0.13	0.35	0.14	0.4
3	5.3	313.7	0.1	0.25	0.14	0.34
4	5.3	329.8	0.1	0.27	0.17	0.45
5	5	373.4	0.13	0.33	0.16	0.41
6	4.3	329.3	0.09	0.24	0.14	0.37
7	4.3	370.8	0.12	0.32	0.14	0.37
8	4.4	341.8	0.11	0.28	0.17	0.46
9	4.9	307.7	0.11	0.28	0.18	0.46
10	3.6	351	0.09	0.21	0.18	0.43
Average	4.7	340.4	0.11	0.28	0.16	0.41
CoV (%)	11.80	7.00	13.20	15.10	10.40	10.00

contributions from both bending and shear deflections of a DCB. The application of such procedure to an anisotropic material like wood can be done from Eq. (3.5) and by applying the Castigliano theorem (de Moura and Dourado, 2018)

$$C = \frac{1}{P}\frac{\partial U}{\partial P} = \frac{24}{E_L B}\int_0^a \frac{x^2}{h^3}dx + \frac{12}{5BG_{LT}}\int_0^a \frac{1}{h}dx \tag{3.20}$$

The dC/da relation becomes

$$\frac{dC}{da} = \frac{12}{B}\left(\frac{2a^2}{E_L h^3} + \frac{1}{5G_{LT}h}\right) \tag{3.21}$$

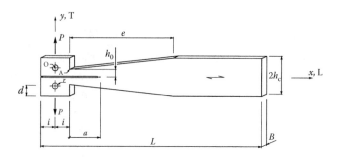

FIGURE 3.25
Configuration of the TDCB specimen with dimensions (in mm): $L = 300$, $B = 15$, $e = 140$, $h_0 = 10$, $h_c = 25$, $i = 20$, $r = 14$ and $d = 12.5$. (Adapted from de Moura and Dourado, 2018).

The main goal is to define a specimen profile so that the term inside brackets in Eq. (3.21) becomes a constant (m), thus leading to the following quite simple relation for strain energy release rate,

$$G_I = \frac{6P^2}{B^2} m \qquad (3.22)$$

Although a curved specimen profile results from the earlier condition, the required shape can be approximated by a linear taper when the ratio a/h is large (Mostovoy et al., 1967). To confirm this statement when applied to wood fracture tests, a numerical analysis (Figure 3.26) was performed considering the specimens dimensions presented in Figure 3.25 using the elastic properties of *Pinus pinaster* Ait. (Table 3.1).

The method used to obtain the fracture energy is based on compliance calibration as a function of the crack length. An automatic procedure providing a fine analysis, i.e. evolution of compliance for very small variations of crack length was considered. Normalised quantities were used in Figure 3.27 aiming the generalisation of these results to other specimen dimensions (i.e. homothetic specimens). It can be observed that an almost linear trend exists when the crack propagates inside the tapered region (i.e. $a/e \leq 1$), which confirms the statement of an almost constant $C = f(a)$ relation.

3.3.3 Compliance-Based Beam Method

The parameter m [term inside brackets in Eq. (3.21)] depends on the used elastic properties, which reveal remarkable variability between specimens. In addition, because of inhomogeneity of wood material, it is common to observe variations in compliance as damage propagates under pure mode I loading. These aspects can induce errors on the measured toughness following the procedure described in the previous section. Therefore, an

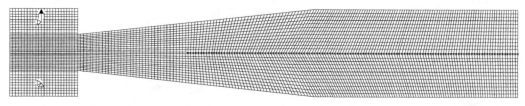

FIGURE 3.26
Mesh used for TDCB specimen.

Mode I Fracture Characterisation

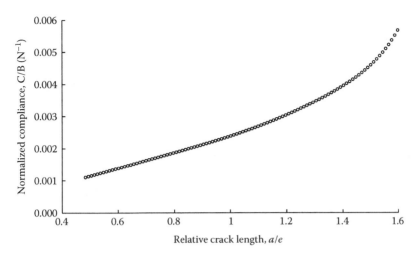

FIGURE 3.27
Normalised compliance versus normalised crack length (see Figure 3.25).

alternative is to employ an equivalent crack length procedure similar to the one described for DCB tests (Section 3.1.4), i.e. the CBBM. In this case (TDCB specimen), the following relations should be established (see Figure 3.25),

$$h(x) = h_0 + C_1 x \quad \text{with } C_1 = (h_c - h_0)/e \tag{3.23}$$

In addition, the loaded segment (head of the specimen) is assumed as a rigid body (i.e. non-deformable). In this context, a constant bending moment ($M_A = Pi$, see Figure 3.25) has to be considered at the beginning of the tapered zone and Eq. (3.20) becomes,

$$C = \frac{1}{P}\frac{\partial U}{\partial P} = \frac{24}{E_L B}\int_0^a \frac{x^2 + i^2}{h^3}dx + \frac{12}{5BG_{LT}}\int_0^a \frac{1}{h}dx \tag{3.24}$$

which leads to

$$C = \frac{24}{E_L B C_1}\left[-\frac{a^2}{2(h_0 + C_1 a)^2} - \frac{a}{C_1(h_0 + C_1 a)} + \frac{\ln(h_0 + C_1 a)}{C_1^2} - \frac{\ln h_0}{C_1^2}\right.$$

$$\left. + \frac{i^2}{2}\left(\frac{1}{h_0^2} - \frac{1}{(h_0 + C_1 a)^2}\right)\right] + \frac{12}{5BG_{LT}C_1}\ln\left(\frac{h_0 + C_1 a}{h_0}\right) \tag{3.25}$$

To account for scatter of elastic properties between specimens, an effective flexural modulus (E_f) can be easily obtained from Eq. (3.25), considering the initial conditions, C_0 and a_0. During propagation, Eq. (3.25) can be solved to provide an equivalent crack length a_e function of the current compliance registered during the fracture test. The solver option of Excel was used to iteratively solve Eq. (3.25) to a_e, since an analytical solution is not available. Applying the Irwin–Kies expression (Eq. 2.39) the R-curve $(G_I = f(a_e))$ is given by,

$$G_I = \frac{6P^2}{B^2(h_0 + C_1 a_e)}\left(\frac{2(a_e^2 + i^2)}{E_f(h_0 + C_1 a_e)^2} + \frac{1}{5G_{LT}}\right) \tag{3.26}$$

which is a result similar to the one obtained for DCB test (Eq. 3.14), with the difference that h is not constant.

The described procedure is based on compliance monitored at the beginning of the tapered zone (point A in Figure 3.25). However, loading is applied at the centre of the specimen head (point O) assumed to be a rigid body. Hence, a corrected displacement must be determined,

$$\delta_A = \delta_O - 2i\sin\theta_A \tag{3.27}$$

where θ_A is the beam rotation at point A, given by

$$\theta_A = \frac{\partial U}{\partial M_A} = \frac{\partial}{\partial M_A}\int_0^a \frac{(M_A + Px)^2}{2E_f I}dx \quad \text{where } M_A = Pi \tag{3.28}$$

After some algebraic manipulation,

$$\theta_A = \frac{12}{E_f B}\left[\frac{M_A}{2C_1}\left(\frac{1}{h_0^2} - \frac{1}{(C_1 + h_0)^2}\right) + \frac{P}{2C_1}\left(-\frac{a}{(C_1 + h_0)^2} - \frac{1}{C_1(C_1 + h_0)} + \frac{1}{C_1 h_0}\right)\right] \tag{3.29}$$

Following this procedure, the compliance at point A is determined and used in Eq. (3.25) to estimate the equivalent crack length that is subsequently utilised in Eq. (3.26) to get the R-curve.

3.3.4 Numerical Validation

A numerical analysis including CZM was performed considering the elastic and fracture properties of Table 3.1, the bilinear softening law (Figure 3.8) and the geometry presented in Figure 3.25 with an initial crack length (a_0) of 80 mm. The evolution of compliance in the course of the test simulation is used to estimate the equivalent crack length a_e.

Some variation is observed for parameter m for the selected tapered geometry (Figure 3.28). Notwithstanding this aspect, a satisfactory linear approach is achieved for $C = f(a_e)$, as observed in Figure 3.29.

The R-curves ensuing from three methods are plotted in Figure 3.30: the CBBM, the method proposed by Mostovoy and the Irwin–Kies expression (Eq. 2.39). It can be affirmed that CBBM method provides good agreement of G_I with the value used as input (i.e. G_{Ic}) in the plateau region. Nevertheless, the Mostovoy and Irwin–Kies based methods slightly

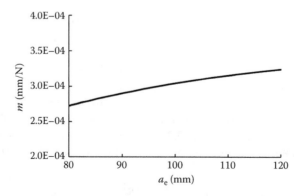

FIGURE 3.28
Evolution of parameter m as a function of crack length.

Mode I Fracture Characterisation

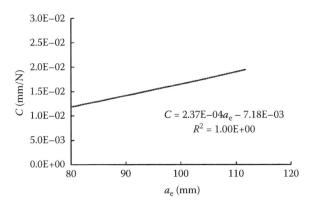

FIGURE 3.29
Compliance versus equivalent crack length relationship.

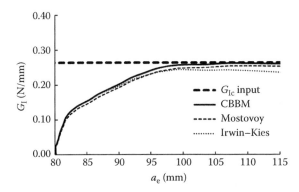

FIGURE 3.30
Numerical *R*-curves of TDCB specimen using the CBBM and Mostovoy-based methods.

underestimate the input value, which can be explained by the fact that the selected specimen tapering does not comply in full with the statement of a constant *m*.

3.3.5 Experimental and Numerical Results

A set of six TDCB specimens with the geometry shown in Figure 3.25 were tested experimentally (Figure 3.31). An initial pre-crack $a_0 = 66$ mm was introduced, thus giving rise to approximately 70 mm of propagation in the tapered region.

FIGURE 3.31
The TDCB test.

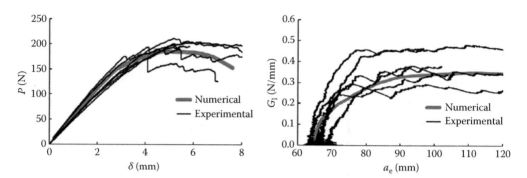

FIGURE 3.32
Experimental and numerical load–displacement and R-curves of six TDCB specimens.

The corresponding load–displacement and R-curves, obtained using the CBBM described in Section 3.3.3, are plotted in Figure 3.32. In the plateau region, the values of toughness are roughly in the range 0.3–0.4 N/mm, thus pointing to an average of $G_{Ic} = 0.35$ N/mm.

For the sake of verification, a numerical analysis was performed considering the bilinear softening law described in Section 3.1.5 with $G_{Ic} = 0.35$ N/mm. The ensuing load–displacement and R-curves were included in Figure 3.32. It can be concluded that both curves are representative of the global experimental trends, thus confirming the validity of the developed procedure.

3.4 Compact Tension Test

The CT test was developed for fracture mechanics tests in metals. Because of its simplicity, it was also adopted for fracture characterisation of wood under mode I loading. The CT test is similar to the DCB one (Figure 3.3a), differing only on the specimen size. In fact, the CT specimen has a shorter length and higher height (Figure 3.33) than the DCB.

The application of the CT test in the context of wood fracture characterisation reveals some difficulties. One of them is the impracticality of applying beam theory based methods for

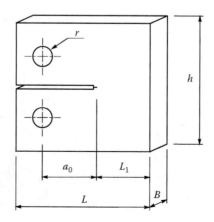

FIGURE 3.33
Schematic representation of the CT specimen: $h = 70, a_0 = 10, B = 20$ mm; $L = 1.25(a_0 + L_1), r = 0.125(a_0 + L_1)$.

data reduction purposes. The viable alternative is using the CCM to establish the $C = f(a)$ relation. Experimentally, this can be done using two different ways. One of them consists in considering several specimens with different pre-crack lengths and determine the respective initial compliance. The inherent material variability among different specimens, characteristic of natural materials as is the case of wood, can be considered an important drawback of this method, since erroneous estimations of toughness will certainly occur. Alternately, the crack length can be monitored in the course of the CT fracture test, which, as already discussed in the context of the DCB tests, is very difficult to perform with the required accuracy.

Another drawback of the CT test in the context of wood fracture characterisation is related with the typical material softening behaviour due to several toughening mechanisms acting during the fracture process. In fact, the quasi-brittle behaviour of wood leads to the development of a non-negligible FPZ that is incompatible with the short ligament length (L_1 in Figure 3.33) characteristic of the CT specimen.

A FEA considering CZM of the CT test was carried out only varying the ligament length (L_1) with the purpose of assessing the mentioned malfunction. The properties and cohesive law used in this analysis were the ones determined in the DCB testing campaign (Table 3.1). The ligament length varied between 20 and 50 mm with a step of 10 mm, and the remaining specimen dimensions are listed in Figure 3.33. It was observed that the profile of the load–displacement curves depends on the ligament length but tend to converge as it increases (Figure 3.34). It should be noted that similar conclusions were obtained by Boström (1994). This author observed that the maximum normalised force depends on the specimen size and concluded that special care should be taken when the maximum load is used for toughness measurement purposes.

To better understand this behaviour, the evolution of the CZL as a function of applied displacement was also investigated (Figure 3.35). In all cases, the CZL increases almost linearly from the beginning of the test till a limit value is attained. For $L_1 = 20$ mm the maximum value is lower relatively to remaining cases and a constant trend (i.e. a plateau) is not attained. This leads to the conclusion that the ligament length of 20 mm is clearly insufficient for a complete development of the CZL, which means that conditions for self-similar crack growth are not fulfilled. In the remaining cases, similar plateau values are obtained but with different extents. In fact, a quite short plateau is obtained for $L_1 = 30$ mm, which increases successively for $L_1 = 40$ mm and 50 mm. This behaviour means that self-similar crack growth conditions are hardly satisfied for $L_1 = 30$ mm but reasonably accomplished for the other two

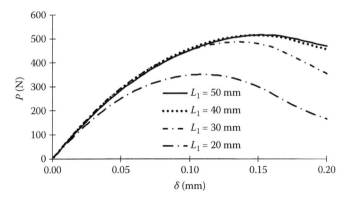

FIGURE 3.34
Load–displacement curves considering different ligament lengths (L_1).

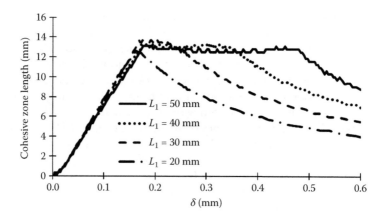

FIGURE 3.35
Evolution of the cohesive zone length as a function of applied displacement considering different ligament lengths (L_1).

cases. This phenomenon also explains the difference between the load–displacement curves obtained for $L_1 = 30$ mm and the other ones that have converged (Figure 3.34).

In summary, it can be concluded that the classical CT test characterised by small ligament lengths is not suitable for wood fracture analyses under mode I loading. It is necessary to guarantee a minimum ligament length that avoids the interference of the natural development of the FPZ with the specimen edge. The size of the necessary ligament length will depend on the wood species and its properties. Owing to similarity among tests, the DCB test is preferable since a quite longer ligament length exists in this specimen.

3.5 Conclusions of Mode I Fracture Tests

Fracture characterisation of wood under mode I loading was thoroughly discussed in this chapter. Four experimental tests were described in detail: the DCB, the SEN-TPB, the TDCB and the CT. Wood reveals to be a quasi-brittle material developing a non-negligible FPZ, which makes unsuitable the application of classical reduction schemes. To overcome such limitation, new data reduction schemes based on Timoshenko beam theory, specimen compliance and crack equivalent concept are proposed for DCB, SEN-TPB and TDCB tests, thus contributing to increase their appropriateness for wood fracture characterisation under mode I loading. The developed methods were validated numerically by means of FEA, including CZM. The analysis of CT test revealed some limitations of this test to provide accurate evaluation of mode I toughness in wood owing to its small ligament length.

References

Boström, L. (1994). The stress-displacement relation of wood perpendicular to the grain—Part 2. Application of the fictitious crack model to the compact tension specimen. *Wood Sci Technol*, 28:319–27.

Coureau, J.-L., S. Morel and N. Dourado (2013). Cohesive zone model and quasibrittle failure of wood: A new light on the adapted specimen geometries for fracture tests. *Eng Fract Mech*, 109:328–40.

Davalos, J. F., P. Madabhusi-Raman, P. Z. Qiao and M. P. Wolcott (1998). Compliance rate change of tapered double cantilever beam specimen with hybrid interface bonds. *Theor Appl Fract Mec*, 29:125–39.

de Moura, M. F. S. F., M. A. L. Silva, A. B. de Morais and J. J. L. Morais (2006). Equivalent crack based mode II fracture characterization of wood. *Eng Fract Mech*, 73:978–93.

de Moura, M. F. S. F., J. J. L. Morais and N. Dourado (2008). A new data reduction scheme for mode I wood fracture characterization using the double cantilever beam test. *Eng Fract Mech* 75:3852–65.

de Moura, M. F. S. F., N. Dourado and J. J. L. Morais (2010). Crack equivalent based method applied to wood fracture characterization using the single edge notched-three point bending test. *Eng Fract Mech*, 77:510–20.

de Moura, M. F. S. F. and N. Dourado (2018). Mode I fracture characterization of wood using the TDCB test. *Theor Appl Fract Mec*, 94:40–5.

Dourado, N., S. Morel, M. F. S. F. de Moura, G. Valentin and J. J. L. Morais (2008). Comparison of fracture properties of two wood species through cohesive crack simulations. *Compos Part A-Appl Sci Manuf*, 39:415–27.

Dourado, N., M. F. S. F. de Moura, J. J. L. Morais and M. A. L. Silva (2010). Estimate of resistance-curve in wood through the double cantilever beam test. *Holzforschung*, 64:119–26.

Dourado, N., M. F. S. F. de Moura, S. Morel and J. J. L. Morais (2015). Wood fracture characterization under mode I loading using the three-point-bending test. Experimental investigation of *Picea abies L. Int J Fract*, 194:1–9.

Gustafsson, P. J. (1988). A study of strength of notched beams. In: Proceedings of CIB-W18A Meeting, Parksville, Canada, Paper 21-10-1.

ISO 15024: 2001 (2001). Fibre-reinforced plastic composites—Determination of mode I interlaminar fracture toughness, GIc, for unidirectionally reinforced materials.

Kienzler, R. and G. Herrmann (1986). An elementary theory of defective beams. *Acta Mech*, 62:37–46.

Morel, S., J. Schmittbuhl, J. M. López and G. Valentin (1998). Anomalous roughening of wood fractured surfaces. *Phys Rev E*, 58:6999–7005.

Morel, S., J. Schmittbuhl, E. Bouchaud and G. Valentin (2000). Scaling of crack surfaces and implications for fracture mechanics. *Phys Rev Lett*, 85:1678–81.

Morel, S., E. Bouchaud, J. Schmittbuhl and G. Valentin (2002). R-curve behavior and roughness development of fracture surfaces. *Int J Fract*, 114:307–25.

Morel, S., G. Mourot and J. Schmittbuhl (2003). Influence of the specimen geometry on R-curve behaviour and roughnening of fracture surfaces. *Int J Fract*, 121:23–42.

Morel, S., N. Dourado, G. Valentin and J. J. L. Morais (2005). Wood: A quasibrittle material—R-curve behavior and peak load evaluation. *Int J Fract*, 131:385–400.

Mostovoy, S., P. B. Crosley and E. J. Ripling (1967). Use of crack-line loaded specimens for measuring plane-strain fracture toughness. *J Mater*, 2:661–8.

Oliveira, J. M. Q., M. F. S. F. de Moura, J. J. L. Morais and M. A. L. Silva (2007). Numerical analysis of the MMB test for mixed-mode I/II wood fracture. *Compos Sci Technol*, 67:1764–71.

4

Mode II Fracture Characterisation

Mode II fracture characterisation of wood has received less of the researchers' attention relative to mode I. Typically, the designers conceive wood structures in order to minimise mode I loading that can induce crack propagation parallel to wood fibres and neglect mode II. However, mode II loading is prone to occur namely in structures under shear or bending loads. In the last case, the mismatch bending existing between earlywood and latewood layers leads to interlaminar shear stresses at the interface that can result in a mode II failure. For this reason and considering the growing interest on wood structural applications, the development of suitable fracture tests becomes relevant, as well as appropriate data reduction schemes to wood fracture characterisation under mode II loading, to accurately predict the material susceptibility to this failure mode.

Few works can be referred about wood fracture characterisation under mode II loading using several tests. Barrett and Foschi (1977) were the first to propose the end-notched flexure specimen (ENF) and applied it to study mode II fracture of western hemlock species. Cramer and Pugel (1982) used a compact shear specimen to determine the mode II fracture toughness of southern pine in the longitudinal-tangential (LT) system. Murphy (1988) proposed a centre-slit beam for the measurement of mode II fracture properties of western hemlock. Kretschmann (1995) investigated the size and thickness effects on the performance of tapered ENF shear specimen, proposed by the International Union of Testing and Research Laboratories for Materials and Structures to establish mode II fracture mechanics properties in wood. Yoshihara and Ohta (2000) examined the validity of the ENF test method to measure the mode II energy release rate of western hemlock, using two distinct calculation methods: the load-loading point compliance method and the load-crack shear displacement (CSD) method. Yoshihara (2004) used the four-point bend ENF (4ENF) test to obtain the *Resistance*-curve (*R*-curve) for spruce wood without measuring the crack length. In general, these works were based on linear elastic fracture mechanics principles. However, there are experimental evidences that wood under mode II loading behaves like a quasi-brittle material, whose fracture is characterised by the development of a large fracture process zone ahead of the crack tip. Consequently, wood mode II fracture behaviour does not follow the linear elastic fracture mechanics predictions, and the non-linear fracture mechanics theory becomes the appropriate theoretical background. In this context, wood fracture can be simulated by cohesive zone models applied through finite element calculations.

In the following, the most common fracture tests for mode II fracture characterisation [i.e. the ENF, the end-loaded split test (ELS) and the 4ENF] are detailed. Specific data reduction schemes based on crack equivalent concepts are described. The experimental toughness results ensuing from these non-classical data reduction schemes are validated numerically by means of cohesive zone models.

67

4.1 End-Notched Flexure Test

The ENF is a simple three-point bending test on a pre-cracked specimen, thus being the most widely used mode II loading, and it was developed for fracture characterisation of wood (Barrett and Foschi, 1977). After that, several authors have studied this test. Russell and Street (1985) developed a solution based on simple beam theory neglecting transverse shear deformation and crack tip singularity. However, their solution was verified to underestimate G_{IIc}. Carlsson et al. (1986) included the transverse shear deformation, but their results also underestimate G_{IIc}. Whitney et al. (1987) incorporated the crack-tip deformation effect in the analysis using the shear deformation plate theory. Wang and Williams (1992) developed a new method to correct the crack length to account for crack tip deformation. The inconvenience of the referred methods is the necessity of crack measurement during propagation, which is very difficult to perform experimentally since the crack grows without a clear opening. To overcome this problem some solutions have been proposed. Yoshihara and Ohta (2000) examined the validity of the ENF testing method to measure wood mode II fracture toughness. The authors recommended the use of CSD to obtain G_{II}, since the crack length is implicitly included in the CSD. The CSD is the relative shear slip between upper and lower crack surfaces of the ENF specimen, which is measured using a special displacement gage (Kageyama et al., 1991). Yoshihara and Ohta (2000) concluded that G_{IIc} depends on the initial crack length. They also verified an increase of G_{II} during crack propagation, until a plateau was reached. Tanaka et al. (1995) concluded that, to extend the stabilised crack propagation range in the ENF test, the test should be done under a condition of controlled CSD. However, this method requires a servo valve-controlled testing machine, and the testing procedure is more complicated than that under the loading point displacement condition. In the following sections, an equivalent crack-based procedure previously developed (de Moura et al., 2006, 2009; Silva et al., 2006) is presented to overwhelm the difficulties inherent to crack length monitoring during the ENF tests.

4.1.1 Test Description

The ENF test consists of loading on a three-point bending fixture a pre-cracked specimen (Figure 4.1) that produces a shear loading in the longitudinal direction at the crack tip, i.e. mode II loading. The load is applied through a cylindrical pin, and the specimen is supported in two cylinders (10 mm diameter).

FIGURE 4.1
The experimental setup applied to ENF test (de Moura et al., 2009).

Mode II Fracture Characterisation

The pre-crack at the specimen mid-plane is executed in two steps: initially a starter notch is introduced by means of a band saw (1 mm thickness); afterwards the crack is extended 2–3 mm by applying a low impact load on a cutting blade. The initial crack length a_0 was accurately measured on both sizes of specimen to confirm its alignment, since it is a fundamental parameter in the considered data reduction methods. The used value (Figure 4.2) satisfies the relation $a_0 > 0.7L$ enabling stable crack growth (Carlsson et al., 1986). Two sheets of Teflon with a pellicle of lubricant between them were placed at the pre-crack to minimise friction effects at this region.

4.1.2 Classical Data Reduction Schemes

There are two main data reduction methods applied to ENF test to measure material toughness, G_{IIc}. The compliance calibration method (CCM) is based on Irwin–Kies equation (2.39). The compliance calibration can be executed by two different ways. One of them consists of performing bending tests using specimens with different initial cracks. This can be achieved using only one specimen by altering its position in the supports. Alternatively, the calibration of C can be reached by measuring the crack length during propagation. A cubic polynomial fitting should be carried out considering

$$C = D + ma^3 \tag{4.1}$$

where D and m are constants. Combining previous equation with Eq. (2.39) yields

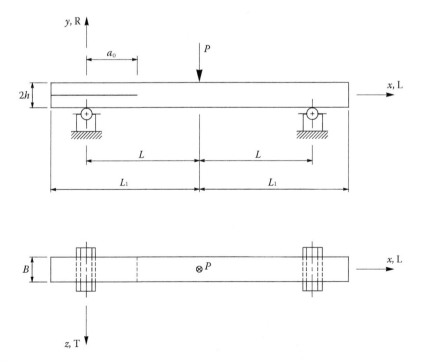

FIGURE 4.2
Schematic representation of the ENF test ($2L_1 = 500$, $2L = 460$, $a_0 = 162$, $2h = 20$, width $B = 20$; all dimensions in mm).

$$G_{II} = \frac{3P^2 m a^2}{2B} \quad (4.2)$$

Alternatively, the beam theory can be used. Wang and Williams (1992) proposed a corrected beam theory (CBT) based on

$$G_{II} = \frac{9(a+0.42\Delta_I)^2 P^2}{16B^2 h^3 E_L} \text{ with } \Delta_I = h\sqrt{\frac{E_L}{11 G_{LR}}\left[3 - 2\left(\frac{\Gamma}{1+\Gamma}\right)^2\right]} \text{ and } \Gamma = 1.18\frac{\sqrt{E_f E_R}}{G_{LR}} \quad (4.3)$$

with $0.42\Delta_I$ being the crack length correction to account for shear deformation. The application of both methods (CCM or CBT) requires crack length measurement during propagation. However, there is a problem intrinsic to all fracture characterisation tests in mode II, which is the remarkable difficulty to rigorously monitor the crack, since it tends to propagate with their faces in contact. In Figure 4.3, a discontinuity can be observed on the black vertical reference line. However, the crack tip position cannot be clearly identified as virtually no changes are visible in the neighbouring material. In fact, it cannot be guaranteed that this discontinuity on the black line corresponds to a clear crack. Owing to pronounced fracture process zone (FPZ) observed under mode II loading in wood, the referred discontinuity can be an indication of the softening process development ahead of the crack tip. In addition, the large FPZ requires that the energy dissipated in it should be accounted for, which reinforces the necessity of a crack equivalent method.

4.1.3 Compliance-Based Beam Method

In the following presentation, the radial-longitudinal (RL) crack propagation system (Figure 4.2) was assumed. The strain energy of the ENF specimen due to bending and including shear effects is

$$U = \int_0^{2L} \frac{M_f^2}{2E_f I} dx + \int_0^{2L} \int_{-h}^{h} \frac{\tau^2}{2G_{LR}} B \, dy \, dx \quad (4.4)$$

where E_f and G_{LR} are the flexural and shear modulus, respectively, M_f is the bending moment and

$$\tau = \frac{3}{2}\frac{V_i}{A_i}\left(1 - \frac{y^2}{c_i^2}\right) \quad (4.5)$$

FIGURE 4.3
Photograph of a crack propagating under pure mode II loading (de Moura et al., 2009).

Mode II Fracture Characterisation

where A_i, c_i and V_i represent, respectively, the cross-section area, half-thickness of the beam and the transverse load of the i segment ($0 \leq x \leq a$, $a \leq x \leq L$ or $L \leq x \leq 2L$). From the Castigliano theorem, the displacement at the loading point for a crack length a is

$$\delta = \frac{dU}{dP} = \frac{P(3a^3 + 2L^3)}{8E_f Bh^3} + \frac{3PL}{10G_{LR}Bh} \tag{4.6}$$

Since the flexural modulus of the specimen plays a fundamental role on the P–δ relationship, it can be estimated from Eq. (4.6) using the initial compliance C_0 and the initial crack length a_0

$$E_f = \frac{3a_0^3 + 2L^3}{8Bh^3}\left(C_0 - \frac{3L}{10G_{LR}Bh}\right)^{-1} \tag{4.7}$$

This procedure takes into account the variation of the material properties between different specimens and several effects that are not included in beam theory, e.g. stress concentration near the crack tip and contact between the two arms. In fact, these phenomena affect the specimen behaviour and consequently the P–δ curve, even in the elastic regime. Using this methodology, their influences are accounted for through the calculated flexural modulus. On the other hand, the effect of a large FPZ that characterises mode II loading should also be considered since it affects the material fracture behaviour. Consequently, during crack propagation, a correction of the real crack length is considered in the compliance Eq. (4.6) to include the FPZ effects

$$C = \frac{3(a + \Delta a_{FPZ})^3 + 2L^3}{8E_f Bh^3} + \frac{3L}{10G_{LR}Bh} \tag{4.8}$$

and consequently,

$$a_e = a + \Delta a_{FPZ} = \left[\frac{C_{corr}}{C_{0corr}}a_0^3 + \frac{2}{3}\left(\frac{C_{corr}}{C_{0corr}} - 1\right)L^3\right]^{1/3} \tag{4.9}$$

where C_{corr} and C_{0corr} are given by

$$C_{corr} = C - \frac{3L}{10G_{LR}Bh}; \quad C_{0corr} = C_0 - \frac{3L}{10G_{LR}Bh} \tag{4.10}$$

G_{IIc} can now be obtained from

$$G_{IIc} = \frac{9P^2 a_e^2}{16B^2 E_f h^3} \tag{4.11}$$

Using this methodology, crack measurements are unnecessary. Experimentally, it is only necessary to register the values of applied load and displacement. The method accounts for FPZ effects through a_e and for the influence of specimen variability on the results by means of the E_f. A typical value for the shear modulus G_{LR} can be used owing to its minor influence on the results relatively to E_f (de Moura et al., 2006). Following this procedure, a complete R-curve [$G_{II} = f(a_e)$] is obtained. The value of G_{IIc} can be estimated from its plateau.

Additionally, since this method provides the *R*-curve, the problems related to unstable crack growth and the spurious increase of G_{II} at crack starting advance due to non-natural pre-crack are mitigated.

4.1.4 Experimental and Numerical Results

The proposed data reduction scheme (CBBM) was applied to experimental tests. Twenty-four specimens of *Pinus pinaster* wood with 12% moisture content and average oven dry specific density equal to 0.55 were machined and tested. The real dimensions of each specimen (initial crack length included) were recorded. Fracture tests were conducted in a displacement control mode using a screw-driven universal testing machine (Instron 1125), considering a crosshead speed of 5 mm/min and an acquisition frequency of 5 Hz. Figure 4.4 presents the load–displacement curves and the corresponding *R*-curves obtained applying the CBBM. Although the samples matched the specimens cut from the same trunk, a remarkable initial stiffness variability can be observed in Figure 4.4a. This aspect reinforces the utility of the CBBM since the initial modulus is a calculated parameter for each specimen (Eq. 4.7). The *R*-curves (Figure 4.4b) present an initial rising trend corresponding to FPZ development followed by a plateau describing self-similar propagation under steady-state conditions. This aspect is fundamental for an effective evaluation of fracture energy under mode II loading, G_{IIc}.

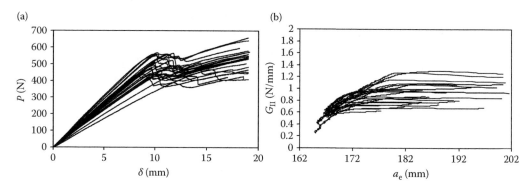

FIGURE 4.4
(a) Load–displacement curves and (b) the corresponding *R*-curves of the ENF test (de Moura et al., 2009).

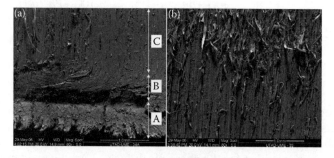

FIGURE 4.5
(a) Initial crack sections: (A) sawn section; (B) sliced section; (C) mode I pre-cracked section. (b) Transition between mode I pre-crack and mode II crack (Silva et al., 2007).

Mode II Fracture Characterisation

Fracture surfaces were observed using a scanning electronic microscope Philips-FEI Quanta 400. Figure 4.5a is a photomicrograph of the initial crack surface, where three different zones are visible. Region A is the tip of the starter notch machined with the band saw, whereas region B is the portion of the initial crack cut by the blade. Zone C of the initial crack corresponds to the mode I fracture surface created ahead of cutting blade, due to its penetration into wood under an impact loading. Indeed, this section exhibits the characteristic appearance of fracture in perpendicular-to-grain tension, through the middle lamella of tracheids (Zink et al., 1994). The transition between the surface of mode I pre-crack and the surface of mode II (parallel to grain shear loading) crack can be clearly seen in Figure 4.5b. The mode II crack surface is characterised by severe twisting, tearing, and unwinding of tracheid walls (Zink et al., 1994).

Numerical analyses considering cohesive zone modelling (CZM) were performed to validate the proposed procedure applied to mode II fracture characterisation. The numerical model includes eight-node isoparametric two-dimensional solid and special developed six-node interface finite elements. The used mesh has 2,750 solid elements and 122 cohesive elements placed at the mid-plane of the uncracked region (Figure 4.6). In the pre-crack region, contact conditions were imposed to prevent interpenetration of the cracked parts. Contacting surfaces were also considered between the specimen and the supports/actuator, which were simulated as rigid bodies. The analyses were performed assuming plane stress conditions and non-linear geometrical behaviour. A loading displacement ($\delta = 10$ mm) was applied incrementally at the specimen mid-span, considering a very small increment value (0.5% of the total displacement) to ensure a smooth propagation process.

Numerical validation was performed considering the simplest triangular cohesive law (Figure 4.7), which is a particular case of the more complex one presented in Section 2.3.2 (Figure 2.16).

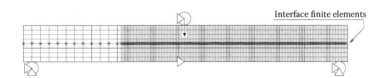

FIGURE 4.6
FE mesh used for the simulation of the ENF test.

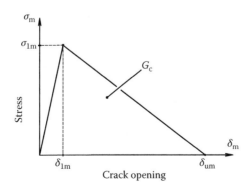

FIGURE 4.7
Triangular cohesive law for mixed-mode I + II loading case.

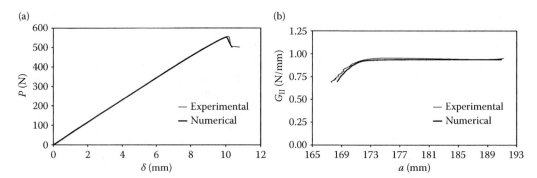

FIGURE 4.8
Experimental and numerical P–δ and R-curves for an ENF test.

Equation (2.48) defines the equivalent displacement leading to damage onset, δ_{1m}. The ultimate relative displacement corresponding to complete failure (δ_{um}) is obtained making $\delta_{3m} = \delta_{1m}$, $\delta_{2m} = \delta_{1m}$ and $\sigma_{3m} = \sigma_{1m}$ in Eq. (2.52), which leads to

$$\delta_{um} = \frac{2G_{Ic}G_{IIc}(1+\beta^2)}{(G_{IIc} + \beta^2 G_{Ic})k\delta_{1m}} \tag{4.12}$$

The damage parameter in this simple case becomes

$$d_m = \frac{\delta_{um}(\delta_m - \delta_{1m})}{\delta_m(\delta_{um} - \delta_{1m})} \tag{4.13}$$

In the numerical validation procedure, all the specimens were simulated considering the G_{IIc} value given by the plateau of the R-curves as an input parameter in the CZM. The local strength was evaluated by an inverse procedure involving the fitting between the numerical and experimental P–δ curves. Figure 4.8 and Table 4.1 reveal the excellent agreement obtained between numerical and experimental results, which validate the ENF test and CBBM as valuable options concerning wood fracture characterisation under mode II loading.

4.2 End-Loaded Split Test (ELS)

The ELS test consists of a cantilever beam with a pre-crack loaded at its extremity to induce crack propagation under pure mode II loading (Figure 4.9). Wang and Vu-Khanh (1996) argue that this method is the most suitable for measuring the R-curve, which is justified by the longer beam length for crack extension that provides more stable fracture condition. Blackman et al. (2005) affirmed that ELS test induces more stable crack propagation and that it is more adequate for toughened adherends concerning their deformations, which must remain elastic during the test. In summary, they concluded that although the ELS test involves more complexities during experiments relative to the ENF test, it provides a larger range of crack length where the crack propagates stably. In fact, the ENF test requires that $a_0/L > 0.7$ to obtain stable crack propagation (Carlsson et al., 1986), whereas in the ELS test

Mode II Fracture Characterisation

TABLE 4.1

Numerical and Experimental Results for All the ENF Tests

Specimen	Experimental		Numerical		Error	
	P_{max} (N)	G_{IIc} (N/mm)	P_{max} (N)	G_{IIc} (N/mm)	P_{max} (%)	G_{IIc} (%)
1	467.4	1.283	488.5	1.277	4.52	−0.50
2	507.5	1.269	528.4	1.264	4.11	−0.33
3	564.4	1.139	579.4	1.143	2.67	0.44
4	469.2	0.772	475.7	0.762	1.39	−1.22
5	414.4	0.667	416.3	0.649	0.48	−2.64
6	441.4	0.721	442.3	0.715	0.22	−0.84
7	549.5	0.994	579.4	0.967	5.44	−2.79
8	493.7	1.077	518.1	1.078	4.95	0.14
9	510.8	0.845	510.3	0.858	−0.11	1.52
10	463.6	0.816	464.2	0.821	0.13	0.64
11	483.5	0.840	492.9	0.817	1.96	−2.67
12	423.0	0.601	425.7	0.590	0.63	−1.80
13	536.2	0.955	536.3	0.941	0.02	−1.44
14	521.6	1.060	536.8	1.052	2.90	−0.74
15	438.4	0.665	440.0	0.649	0.38	−2.40
16	567.7	0.992	566.6	0.980	−0.20	−1.21
17	441.0	0.845	459.3	0.833	3.93	−1.39
18	480.8	0.960	489.6	0.956	1.82	−0.40
19	556.6	0.946	552.4	0.936	−0.74	−1.02
20	474.6	0.767	472.2	0.759	−0.50	−1.06
21	442.9	0.721	441.1	0.708	−0.41	−1.71
22	558.8	0.934	562.6	0.922	0.68	−1.36
23	559.3	1.085	576.5	1.075	3.08	−0.87
24	493.6	0.854	510.3	0.833	3.39	−2.50
Average	494.2	0.91	502.7	0.90	1.86	1.32
Coefficient of Variation (CoV) (%)	9.83	20.16	10.24	20.63		

Source: Adapted from de Moura et al. (2009).

$a_0/L > 0.55$ (Wang and Vu-Khanh, 1996) is sufficient. Silva et al. (2007) performed an extensive analysis of the ELS test when applied to fracture characterisation under pure mode II loading of *Pinus pinaster* wood in the RL system (Figure 4.10), which is the most frequent crack propagation system.

4.2.1 Test Description

In the ELS test, the load is applied near to specimen extremity through a cylindrical pin, and the test fixture includes a linear guidance system allowing horizontal movement of translation of the clamping grip during loading (Figure 4.10). The initial crack length, $a_0 = 105$ mm, was chosen to provide a stable crack propagation, i.e. $a_0/L > 0.55$ (Wang and Vu-Khanh, 1996). The pre-crack execution and measurement were performed according to the procedure described in the ENF tests (Section 4.1.1). Friction effects at the pre-crack were minimised by insertion of two sheets of Teflon with a pellicle of lubrication between them.

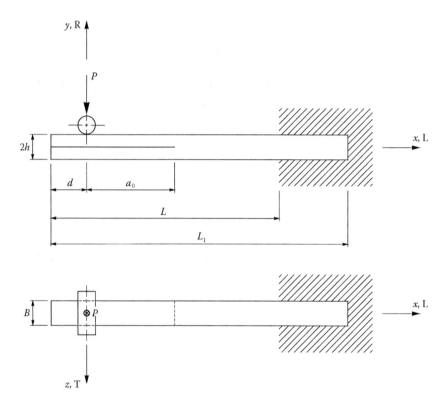

FIGURE 4.9
Schematic representation of the ELS test with specimen nominal dimensions (in mm): $L_1 = 230$, $L = 175$, $2h = 20$, $d = 25$, $B = 20$, and $a_0 = 105$.

FIGURE 4.10
The experimental setup of the ELS test (Silva et al., 2007).

4.2.2 Classical Data Reduction Schemes

The application of CCM to ELS is similar to the one described for ENF, i.e. all the procedure and Eqs. (4.1) and (4.2) can be used. The CBT is based on the following equation

$$G_{\mathrm{IIc}} = \frac{9(a + 0.49\Delta_\mathrm{I})^2 P^2}{4B^2 h^3 E_\mathrm{L}} \qquad (4.14)$$

Mode II Fracture Characterisation 77

where $0.49\Delta_I$ is the crack length correction to account for shear deformation, with Δ_I being given in Eq. (4.3). As occurred in the ENF test, the clear identification of the crack tip during propagation is difficult, since crack grows with their faces in contact. This aspect makes these classical data reduction schemes inaccurate and justifies the development of a crack equivalent based method for the ELS test.

4.2.3 Compliance-Based Beam Method

The development of CBBM for the ELS test is somewhat different relative to ENF test. In fact, the ELS has one additional source of variability that is the clamping conditions that are never perfect as assumed in the beam theory based formulation. The initial conditions (a_0 and C_0) are used to correct the effective length of the specimen, defined as being the one that should be considered to satisfy the beam theory assumptions. Consequently, the longitudinal modulus E_L must be measured for each specimen before doing the ELS test, to avoid the spurious influence of elastic properties variableness on the measured toughness.

Following the procedure described in Section 4.1.3, the compliance equation for the ELS test is

$$C = \frac{3a^3 + L^3}{2Bh^3E_L} + \frac{3L}{5BhG_{LR}} \tag{4.15}$$

For the initial conditions a_0 and C_0, it can be written

$$C_0 - \frac{3a_0^3}{2Bh^3E_L} = \frac{L_{ef}^3}{2Bh^3E_L} + \frac{3L_{ef}}{5BhG_{LR}} \tag{4.16}$$

with L_{ef} being the effective length on the condition of perfect clamping. During propagation, Eq. (4.15) can be rewritten as

$$C - \frac{3a_e^3}{2Bh^3E_L} = \frac{L_{ef}^3}{2Bh^3E_L} + \frac{3L_{ef}}{5BhG_{LR}} \tag{4.17}$$

which, combined with Eq. (4.16), yields the equivalent crack length a_e

$$a_e = \left[(C - C_0)\frac{2Bh^3E_L}{3} + a_0^3 \right]^{1/3} \tag{4.18}$$

The evolution of fracture energy as a function of equivalent crack length (R-curve) can be obtained using the Irwin–Kies relation (Eq. 2.39)

$$G_{II} = \frac{9P^2}{4B^2h^3E_L} \left[(C - C_0)\frac{2Bh^3E_L}{3} + a_0^3 \right]^{2/3} \tag{4.19}$$

Following this methodology, the monitoring of the crack length in the course of the test is not necessary, since an equivalent crack length is defined from current compliance. Moreover, the equivalent crack length and fracture energy do not depend on the effective specimen length (L_{ef}) which is advantageous. However, both depend on E_L that must be previously measured for each specimen in a three-point bending test, owing to its inherent variability typical of a natural material.

4.2.4 Experimental and Numerical Results

Nineteen specimens of *Pinus pinaster* wood with 12% moisture content and average oven dry specific density equal to 0.57 were machined. The real dimensions of each specimen (initial crack length included) were recorded. Three-point bending tests were executed to measure the longitudinal modulus. These tests were performed on specimens with a length of 500 mm and a span equal to 460 mm. Subsequently, these specimens were cut and used to carry out the ELS fracture tests that were performed in a displacement control mode using a screw-driven universal testing machine (Instron 1125), considering a crosshead speed of 5 mm/min and an acquisition frequency of 5 Hz.

Load–displacement curves and the corresponding *R*-curves are presented in Figure 4.11. The effect of material variability is clearly visible in both types of curves. All the *R*-curves reveal a clear plateau corresponding to self-similar crack growth for a given crack extent, which provides an adequate evaluation of G_{IIc} and demonstrates the appropriateness of the ELS test for mode II fracture characterisation of wood.

A numerical analysis involving CZM was also performed (Figure 4.12) to validate the procedure developed for the ELS test. The two-dimensional numerical model includes 1,211 eight-node isoparametric plane stress elements and 130 compatible six-node cohesive elements along the crack path. At the initial crack, 'opened' cohesive elements were considered to allow free relative sliding of the crack surfaces (friction effects were neglected owing to the use of two sheets of Teflon with a pellicle of lubrication between them at the pre-crack region) and prevent interpenetration of the specimen arms. Contact conditions were imposed between the actuator assumed as a rigid body and the specimen.

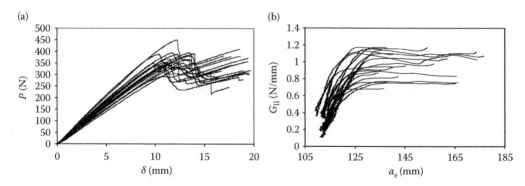

FIGURE 4.11
(a) Load–displacement curves and (b) the corresponding *R*-curves of the ELS test (Silva et al., 2007).

FIGURE 4.12
FE mesh used for the simulation of the ELS test.

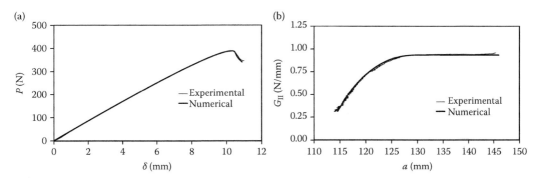

FIGURE 4.13
Experimental and numerical (a) load-displacement curves and (b) the corresponding R-curves of an ELS test.

A geometrically non-linear analysis considering small increment values (0.5% of the applied displacement) was performed to obtain smooth crack propagation.

To validate the CBBM, numerical simulations considering the linear cohesive law were performed for each specimen using their measured dimensions of cross section, initial crack length, longitudinal modulus and G_{IIc}. The effective length (L_{ef}) of each specimen, given by Eq. (4.16), was used in the numerical model instead of the real length of the beam outside the clamping fixture (dimension L in Figure 4.9). This allows the consideration of a perfect clamping condition on the numerical model. A fine-tune iterative process was performed for each specimen to determine the local strength (σ_{1LR}) by means of the minimisation of the difference between the numerical and experimental peak load values. An average value of $\sigma_{1LR} = 9.27$ MPa with a standard deviation of 2.82 MPa was found. Figure 4.13 shows the comparison between numerical and experimental results for a given specimen, and Table 4.2 reveals the global scenario. The excellent agreement observed proves the adequacy of the ELS and the associated procedure concerning wood fracture characterisation under pure mode II loading.

As a final remark, it should be emphasised that the average G_{IIc} resulting from 19 ELS tests ($G_{IIc} = 0.94$ N/mm) is consistent with the one given by the 24 ENF tests ($G_{IIc} = 0.91$ N/mm). This important result demonstrates that experimental campaigns with a large number of specimens allow to overcome the disadvantage of material variability typical of a natural material.

4.3 Four End-Notched Flexure Test

The 4ENF test consists of loading a pre-cracked specimen in four-point bending (Figure 4.14) instead of three-point bending as in the case of ENF. The advantage of 4ENF is related to a constant bending moment between the specimen loading points, which leads to toughness being independent of crack length. Yoshihara (2004) conducted 4ENF fracture tests on spruce and determined its toughness using beam theory and CCMs. The beam theory method allows to determine the fracture toughness without measuring crack length during propagation, although the Young's modulus should be measured in separate tests. In the CCM, specimens with different crack lengths are used to establish the compliance–crack length relation, which is regressed by a linear function. An alternative method is the calculation of fracture toughness and crack length by the load–strain compliance and the combination of load-loading line and load–strain compliances, respectively. This method

TABLE 4.2
Numerical and Experimental Results for All the ELS Tests

	Experimental		Numerical		Error	
Specimen	P_{max} (N)	G_{IIc} (N/mm)	P_{max} (N)	G_{IIc} (N/mm)	P_{max} (%)	G_{IIc} (%)
1	387.8	1.04	390.1	1.03	0.58	−1.05
2	338.3	0.76	337.3	0.75	−0.29	−2.3
3	394.9	1.08	392.2	1.08	−0.7	0.11
4	344.8	0.82	337.9	0.81	−1.98	−1.04
5	449.8	1.15	461.1	1.14	2.51	−0.53
6	349.4	1.16	354.6	1.15	1.48	−1.01
7	362.2	1.11	354.6	1.09	−2.1	−1.66
8	349.6	0.94	348.1	0.92	−0.41	−1.98
9	390.7	1.15	384.4	1.14	−1.63	−0.91
10	342.2	0.86	344.4	0.85	0.63	−1.08
11	350.5	0.82	354.1	0.81	1.02	−0.58
12	352.2	0.91	352.2	0.9	0.01	−0.68
13	388	0.92	389.1	0.91	0.3	−0.98
14	292.9	0.73	298.4	0.73	1.87	0.19
15	381.2	1.09	385.3	1.08	1.06	−0.78
16	394.4	1.03	391.7	1.03	−0.71	−0.66
17	368.5	0.84	361.7	0.85	−1.86	0.66
18	329.6	0.75	331.2	0.75	0.48	−0.89
19	349.3	0.68	342.6	0.67	−1.92	−1.69
Average	364	0.94	363.7	0.93		
CoV (%)	9.2	16.9	9.5	16.9		

Source: Adapted from Silva et al. (2007).

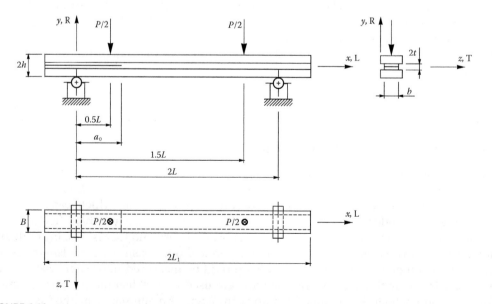

FIGURE 4.14
Schematic representation of the 4ENF test with specimen nominal dimensions (in mm): $2L_1 = 500$, $2L = 450$, $2h = 15$, $2t = 2$, $B = 20$, $b = 10$, $a_0 = 155$.

leads to more complex experimental setup but avoids the need of separate tests to evaluate the longitudinal Young's modulus or dC/da, which characterises the other referred methods. However, the 4ENF specimens with rectangular cross-section usually undergo bending failure of the cracked portion near the roller loading before crack propagation. It was necessary to prepare specimens with an I-shaped cross section to prevent the phenomenon of bending failure and to induce mode II fracture. The author concluded that the proposed methodology is promising to determine the *R*-curve of wood, without monitoring the crack length and without conducting separate tests to evaluate elastic properties.

4.3.1 Test Description

The 4ENF test consists of loading a pre-cracked specimen on a four-point bending special fixture (Figure 4.15). The special device is constituted by cylindrical supports and actuators (10 mm diameter). The upper loading jig is free to rotate around the central pin connecting with machine to assure equal loading forces at the contacts with the specimen. Preliminary tests revealed that premature failure occurred in the vicinity of loading points. To overcome this drawback, two longitudinal grooves were machined in the 4ENF specimens (I section) to diminish the resistant section at the pre-crack plane. The procedure for pre-crack execution and measurement was similar to the one followed for the ENF test. According to Schuecker and Davidson (2000), friction effects in the 4ENF test are more important than in the ENF one. To minimise them at the pre-crack region, two sheets of Teflon with a pellicle of lubrication between them were placed at the pre-crack region.

4.3.2 Compliance Calibration Method

In the case of the 4ENF test, the specimen compliance versus crack length $[C = f(a)]$ relationship is defined by a linear function when *a* propagates between the two loading points

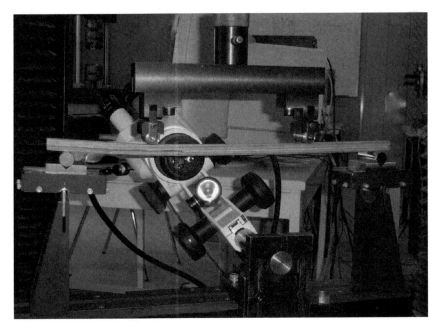

FIGURE 4.15
The experimental setup applied to 4ENF test.

$$C = C_1 a + C_2 \tag{4.20}$$

where C_1 and C_2 are the fitting coefficients. The CCM requires the combination of previous equation with Eq. (2.39) yielding

$$G_{II} = \frac{C_1 P^2}{2B} \tag{4.21}$$

From this equation, it can be concluded that toughness does not depend on the crack length during propagation although the compliance calibration does it. Since crack monitoring in mode II fracture tests is difficult, the calibration can be performed considering different initial pre-crack lengths. The same specimen can be used altering its position in the supports to have different initial crack lengths.

4.3.3 Compliance-Based Beam Method

The CBBM also does not require crack length monitoring in the course of the test and allows overcoming the requirement of compliance calibration of the CCM. According to free body diagram of Figure 4.16, the reaction forces

$$R_A = \frac{P_2 d + P_1 (2L - d)}{2L}; \quad R_B = \frac{P_1 d + P_2 (2L - d)}{2L} \tag{4.22}$$

and bending moments

$$
\begin{aligned}
0 \leq x \leq d &\Rightarrow M_{f1} = \frac{R_A x}{2} \\
d \leq x \leq a &\Rightarrow M_{f2} = \frac{(P_2 - P_1)d}{4L} x + \frac{P_1 d}{2} \\
a \leq x \leq 2L - d &\Rightarrow M_{f3} = \frac{(P_2 - P_1)d}{2L} x + P_1 d \\
2L - d \leq x \leq 2L &\Rightarrow M_{f4} = \left[P_2 (d - 2L) - P_1 d \right] \frac{x - 2L}{2L}
\end{aligned}
\tag{4.23}
$$

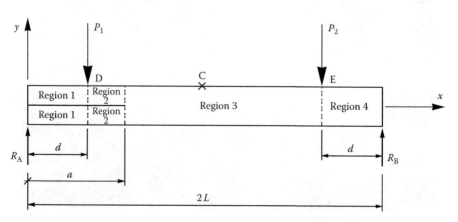

FIGURE 4.16
Free body diagram of the 4ENF specimen test.

Mode II Fracture Characterisation

can be defined.

The strain energy due to bending effects

$$U = \int_0^{2L} \frac{M_f^2}{2E_f I} dx \quad (4.24)$$

must be evaluated for the different specimen regions (Figure 4.16)

$$U_1 = \frac{\left[P_2 d + P_1(2L-d)\right]^2 d^3}{48L^2 E_f I'}$$

$$U_2 = \frac{1}{4E_f I'} \left[\frac{(P_2-P_1)^2 d^2}{12L^2}(a^3-d^3) + \frac{(P_2-P_1)P_1 d^2}{2L}(a^2-d^2) + P_1^2 d^2(a-d)\right]$$

$$(4.25)$$

$$U_3 = \frac{1}{2E_f I} \left\{\frac{(P_2-P_1)^2 d^2}{12L^2}\left[(2L-d)^3 - a^3\right] + \frac{(P_2-P_1)P_1 d^2}{2L}\left[(2L-d)^2 - a^2\right] + P_1^2 d^2(2L-a-d)\right\}$$

$$U_4 = \frac{\left[P_2(d-2L) - P_1 d\right]^2}{24 E_f L^2 I} d^3$$

where I and I' represent the second moment of area of the entire and half specimen section (Figure 4.17), respectively,

$$I = \frac{2\left[Bh^3 - (B-b)t^3\right]}{3}; \quad I' = \frac{Be_1^3 + be_2^3 + (B-b)(e_2-t)^3}{3} \quad (4.26)$$

where

$$e_1 = \frac{bh^2 + (B-b)(h-t)^2}{2(B(h-t)+bt)}; \quad e_2 = h - \frac{bh^2 + (B-b)(h-t)^2}{2(B(h-t)+bt)} \quad (4.27)$$

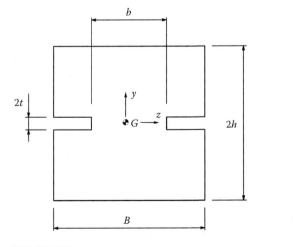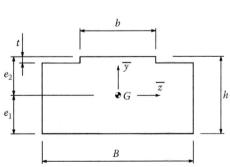

FIGURE 4.17
Entire and half specimen sections.

Applying the Castigliano theorem, the displacements at points D and E (Figure 4.16) are obtained.

$$\delta_D = \frac{dU}{dP_1}; \quad \delta_E = \frac{dU}{dP_2} \tag{4.28}$$

which leads to the displacement of the central point

$$\delta_c = \frac{\delta_D + \delta_E}{2} \tag{4.29}$$

Combining Eqs. (4.25), (4.28) and (4.29) and making $P_1 = P_2 = P/2$ yields

$$\delta_c = \frac{Pd}{2E_f} \left(\frac{-2d^2 + 3da}{12I'} + \frac{-2d^2 - 3da + 6dL}{6I} \right) \tag{4.30}$$

The initial compliance and crack length can be used to estimate an equivalent specimen modulus

$$E_f = \frac{d}{24C_0} \left(\frac{3da_0 - 2d^2}{I'} + \frac{12dL - 4d^2 - 6da_0}{I} \right) \tag{4.31}$$

This parameter allows to account for shear effects and elastic properties variability between different specimens. During propagation, an equivalent crack length accounting for fracture process zone can be defined from Eq. (4.30)

$$a_e = a + \Delta a_{ZPF} = \left(\frac{12E_f C}{d} + \frac{d^2}{I'} - \frac{6dL}{I} + \frac{2d^2}{I} \right) \times \left(\frac{3d}{2I'} - \frac{3d}{I} \right)^{-1} \tag{4.32}$$

Combining the Irwin–Kies relation (Eq. 2.39) and Eq. (4.30) yields

$$G_{II} = \frac{P^2 d^2}{8bE_f} \left(\frac{1}{2I'} - \frac{1}{I} \right) \tag{4.33}$$

As for the CCM, toughness does not depend on the crack length. However, using this compliance beam based method, it is not necessary to measure previously the specimen modulus and perform the compliance calibration with additional tests, and the R-curve [i.e. $G_{II} = f(a_e)$] is easily obtained.

4.3.4 Experimental and Numerical Results

Twenty-five specimens of *Pinus pinaster* Ait. wood with 12% moisture content and average oven dry specific density equal to 0.55 were machined and tested. The real dimensions of each specimen (initial crack length included) were recorded. Fracture tests were conducted in a displacement control mode using a screw-driven universal testing machine (Instron 1125), considering a crosshead speed of 5 mm/min and an acquisition frequency of 5 Hz. Some of the specimens that broke near the loading points before crack growth were not considered, which originated 17 valid results (Figure 4.18). The R-curves reveal a clear plateau allowing to valid evaluation of fracture toughness under mode II loading.

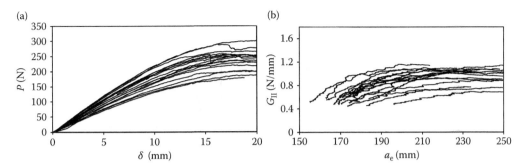

FIGURE 4.18
Load–displacement and the corresponding R-curves of the 4ENF tests.

A finite element analysis of the 4ENF test considering CZM was also performed, aiming to confirm the validity of the procedure concerning mode II fracture characterisation. Owing to longitudinal grooves, a three-dimensional numerical model considering half specimen (longitudinal symmetry conditions) was developed (Figure 4.19). The model is constituted by 35,500 eight-node brick elements and 5,840 eight-node cohesive elements. Contact conditions were imposed between the loading/support cylindrical pins (assumed as rigid bodies) and the specimen. The displacement (δ = 20 mm) was applied to the central cylindrical pin that is linked to the other loading pins by a 'beam-type' connection. The analyses were performed assuming non-linear geometrical behaviour and small increments (0.5% of the total displacement) to assure stable crack growth.

Figure 4.20 shows that the value of G_{IIc} is well captured by the numerical model. The experimental load–displacement curve presents some non-linearity before the peak load that is not well captured by the numerical model. This is probably induced by the

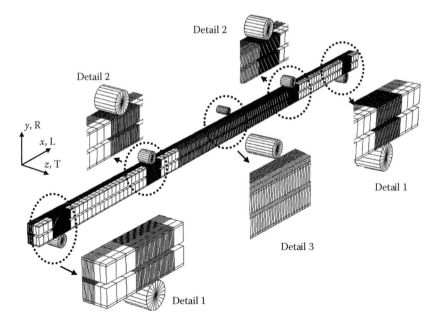

FIGURE 4.19
Three-dimensional finite element model developed for the 4ENF test.

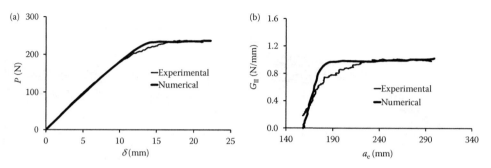

FIGURE 4.20
Numerical and experimental P–δ and R-curves for a 4ENF test.

compliance of the experimental setup that is more complicated in this test. However, the values of G_{IIc} are not influenced by this spurious effect in the region of stable crack growth.

Table 4.3 presents the global results. It can be observed that the global average value (G_{IIc} = 0.97 N/mm) is higher but of the same order to the previous ones given by the ENF and ELS tests. To verify whether the friction effects referred by Schuecker and Davidson (2000) could explain this slight overestimation, two additional finite element analyses were executed considering at the pre-crack region friction coefficients of 0.5 and 1. Overestimations of 3.8% and 4.2% relative to input G_{IIc} were achieved, respectively. Although friction effects were diminished by the introduction of two sheets of Teflon with a lubricant pellicle

TABLE 4.3

Numerical and Experimental Results for All the 4ENF Tests

	Experimental		Numerical		Error	
Specimen	P_{max} (N)	G_{IIc} (N/mm)	P_{max} (N)	G_{IIc} (N/mm)	P_{max} (%)	G_{IIc} (%)
1	218.48	0.913	222.85	0.934	2	2.3
2	254.73	1.064	258.35	1.088	1.42	2.26
3	198.87	0.76	203.34	0.781	2.25	2.76
4	287.64	1.151	289.99	1.158	0.82	0.61
5	252.79	1.056	258.51	1.076	2.26	1.89
6	249.57	1.031	253.84	1.06	1.71	2.81
7	234.57	0.865	239.4	0.89	2.06	2.89
8	232.26	0.983	235.8	1.001	1.52	1.83
9	264.83	1.008	269.94	1.037	1.93	2.88
10	186.07	0.687	192.26	0.704	3.33	2.47
11	177.83	0.767	181.74	0.789	2.2	2.87
12	253.84	1.026	255.5	1.053	0.65	2.63
13	248.79	1.071	251.08	1.079	0.92	0.75
14	229.2	1.086	234.2	1.109	2.18	2.12
15	298.84	1.143	303.69	1.175	1.62	2.8
16	255.83	1.058	258.15	1.081	0.91	2.17
17	199.92	0.902	203.68	0.923	1.88	2.33
Average	237.88	0.97	241.9	1	1.74	2.26
CoV (%)	14.1	14.12	13.72	13.84		

between them, it can be expected that some overestimation of toughness is obtained when using the 4ENF test. The comparison between numerical and experimental values of P_{max} and G_{IIc} reveals a general good agreement that validates the procedure.

4.4 Conclusions of Mode II Fracture Tests

Three types of tests were analysed concerning mode II fracture characterisation of wood: the ENF, the ELS and the 4ENF. Crack equivalent based data reduction schemes were developed for all the tests, which simplifies their execution and provides more accurate evaluations of fracture energy under mode II loading, since the FPZ effects are taken into account. It can be concluded that all of them can be considered valuable options for wood fracture characterisation under mode II loading, as verified in numerical validation procedures involving CZM. However, they present different difficulties of execution. The ELS test requires a special text fixture, allowing horizontal movement of the clamping grip during loading and a pre-measurement of the longitudinal modulus for each specimen. The 4ENF test presents some problems with non-negligible shear effects at the cracked region, execution of longitudinal grooves to avoid premature failure occurring near loading points and requires a special loading device constituted by cylindrical supports and actuators. The ENF test is very simple to execute, does not require any special testing device and the use of the proposed equivalent crack length data reduction scheme minimises the disadvantage of crack instability and crack monitoring. The main outcome of this chapter is that the ENF is the best option for wood fracture characterisation under mode II loading.

References

Barrett, J. D. and R. O. Foschi (1977). Mode II stress-intensity factors for cracked wood beams. *Eng Fract Mech*, 9:371–8.

Blackman, B. R. K., A. J. Kinloch and M. Paraschi (2005). The determination of the mode II adhesive fracture resistance, GIIc, of structural adhesive joints: An effective crack length approach. *Eng Fract Mech*, 72:877–97.

Carlsson, L. A., J. W. Gillespie and R. B. Pipes (1986). On the analysis and design of the End-Notched Flexure (ENF) specimen for mode II testing. *J Compos Mater*, 20:594–604.

Cramer, S. M. and A. D. Pugel (1982). Compact shear specimen for wood mode II fracture investigations. *Int J Fract*, 35:163–74.

de Moura, M. F. S. F., M. A. L. Silva, A. B. de Morais and J. J. L. Morais (2006). Equivalent crack based mode II fracture characterization of wood. *Eng Fract Mech*, 73:978–93.

de Moura, M. F. S. F., M. A. L. Silva, J. J. L. Morais, A. B. de Morais and J. L. Lousada (2009). Data reduction scheme for measuring G_{IIc} of wood in end-notched flexure ENF tests. *Holzforschung*, 63:99–106.

Kageyama, K., M. Kikuchi and N. Yanagisawa (1991). Stabilized end-notched flexure test: Characterization of mode II interlaminar crack growth. ASTM STP 1991;1110:210–25.

Kretschmann, D. E. (1995). Effect of varying dimensions on tapered end-notched flexure shear specimen. *Wood Sci Technol*, 29:287–93.

Murphy, J. F. (1988). Mode II wood test specimen: Beam with center slit. *J Test Eval*, 16:364–8.

Russell, A. J. and K. N. Street (1985). Moisture and temperature effects on the mixed-mode delamination fracture of unidirectional graphite/epoxy. ASTM STP 1985;876:349–70.

Silva, M. A. L., M. F. S. F. de Moura and J. J. L. Morais (2006). Numerical analysis of the ENF test for mode II wood fracture. *Compos Part A-Appl Sci Manuf*, 37:1334–44.

Silva, M. A. L., J. J. L. Morais, M. F. S. F. de Moura and J. L. Lousada (2007). Mode II wood fracture characterization using the ELS test. *Eng Fract Mech*, 74:2133–47.

Schuecker, C. and B. D. Davidson (2000). Evaluation of the accuracy of the four-point bend end-notched flexure test for mode II delamination toughness determination. *Compos Sci Technol*, 60:2137–46.

Tanaka, K., K. Kageyama and M. Hojo (1995). Prestandardization study on mode II interlaminar fracture toughness test for CFRP in Japan. *Composites*, 26:243–55.

Wang, Y. and J. G. Williams (1992). Corrections for mode II fracture toughness specimens of composite materials. *Compos Sci Technol*, 43:251–6.

Wang, H. and T. Vu-Khanh (1996). Use of end-loaded-split (ELS) test to study stable fracture behaviour of composites under mode II loading. *Compos Struct*, 36:71–9.

Whitney, J. M., J. W. Gillespie and L. A. Carlsson (1987). Singularity approach to the analysis of the end-notched flexure specimen. In: Proceedings of the American Society for Composites 2nd Technical Conference, Lancaster, pp. 391–8.

Yoshihara, H. and M. Ohta (2000). Measurement of mode II fracture toughness of wood by the end-notched flexure test. *J Wood Sci*, 46:273–8.

Yoshihara, H. (2004). Mode II R-curve of wood measured by 4-ENF test. *Eng Fract Mech*, 71:2065–77.

Zink, A. G., J. Pellicane and C. E. Shuler (1994). Ultrastructural analysis of softwood fracture surfaces. *Wood Sci Technol*, 28:329–38.

5

Mixed-Mode I + II Fracture Characterisation

In the majority of real applications, structures behave under mixed-mode I + II loading. These mixed-mode conditions arise from external loading and from material anisotropy. The orthotropic directions of wood [longitudinal (L), radial (R) and tangential (T)] present large differences in stiffness and strength. As a result, fracture rarely occurs across the wood fibres, since its strength is much higher than in the radial and tangential directions. Consequently, cracks tend to propagate along the fibre direction independent of its initial orientation, thus inducing mixed-mode fracture conditions that should be accounted for in design, since they can drastically affect the strength predictions. Therefore, it is fundamental to study damage growth under mixed-mode I + II loading to establish adequate fracture criterion. In this context, the development of suitable experimental tests and manageable data reduction schemes for mixed-mode I + II wood fracture characterisation becomes a relevant research topic.

There is no standardised test for mixed-mode I + II fracture characterisation of wood, although several authors have proposed different experimental tests. The double cantilever beam (DCB) with asymmetrical arms and the single edge notched tensile tests are characterised by specimens notched with an inclination relative to the load direction and were used by Jernkvist (2001) to investigate mixed-mode I + II in *Picea abies*. Tschegg et al. (2001) used the asymmetrical wedge splitting test to evaluate the fracture mechanical material properties of spruce under mixed-mode I + II. In this test, the mode I/II ratio can be changed using asymmetrical wedges with different angles. In both cases, fracture was characterised at initiation, and some ambiguity on the definition of critical load was found.

In this chapter, three different tests will be analysed when applied to wood fracture characterisation under mixed-mode I + II loading in the radial-longitudinal (RL) crack propagation system. The single-leg bending (SLB) and the end loaded split for mixed-mode (ELS-MM) tests are similar to end-notched flexure (ENF) and ELS, respectively, which were used for mode II fracture characterisation of wood. They are very simple to execute, but they are limited regarding the variation of the mode ratio. The mixed-mode bending test (MMB) is a combination of the DCB and ENF tests and requires a complex setup. However, the MMB test provides an easy variation of the mode ratio, only altering the length of the loading lever. This is a relevant aspect concerning the definition of a mixed-mode I + II fracture criterion.

The tested material was maritime pine (*Pinus pinaster* Ait.) wood from the central region of Portugal. All specimens were fabricated from boards of 2 m part within the stem's tree. The boards were kiln dried in a standard commercial process, and further conditioned at ambient temperature (20°C–25°C) and 60%–65% relative humidity for at least 4 weeks. The fracture tests were also performed at these environmental conditions. Real dimensions and mass of each specimen were measured, and the density of the tested material was 550 kg m^{-3} for 12.3% moisture content.

5.1 Single-Leg Bending Test

5.1.1 Test Description

The SLB test is a three-point bending test similar to the ENF (Figure 5.1). This test was used in the context of interlaminar fracture characterisation of composites (Szekrényes and Uj, 2004), but never applied to wood. The difference relative to the ENF test is on the used specimen. In the case of the SLB test, the lower specimen arm in the region of the pre-crack is cut (Figure 5.1) to induce mixed-mode I + II (opening and shear modes) instead of pure mode II loading.

To verify an eventual influence of the pre-crack size on the ensuing mode ratio and toughness values, four pre-crack lengths were considered (Figure 5.2). The pre-crack execution and measurement followed the same procedure described for the ENF tests.

FIGURE 5.1
Experimental setup for the SLB test (Oliveira et al., 2009).

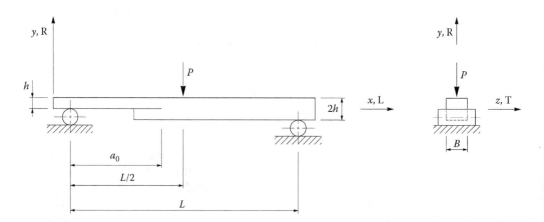

FIGURE 5.2
Schematic representation of the SLB test ($2h = 20$, $L = 480$, $B = 20$ and $a_0 = 168, 183, 198$ and 213; all dimensions in mm).

5.1.2 Compliance-Based Beam Method

The application of the classical data reduction schemes to SLB tests is similar to the already described ENF and is subjected to the same problems already related in pure mode II fracture characterisation tests. The difficulties on crack monitoring, unstable crack growth and extensive fracture process zone (FPZ) justify the development of the compliance based beam method (CBBM) applied to the SLB test. Considering the Timoshenko beam theory, the specimen compliance C becomes (Oliveira et al., 2009)

$$C = \frac{28a^3 + L^3}{32E_L Bh^3} + \frac{3(a+L)}{20G_{LR}Bh} \tag{5.1}$$

An effective specimen modulus can be determined considering the initial conditions (a_0 and C_0) in Eq. (5.1),

$$E_f = \left(C_0 - \frac{3(a_0 + L)}{20G_{LR}Bh} \right)^{-1} \frac{28a_0^3 + L^3}{32Bh^3} \tag{5.2}$$

Using this procedure, the pre-measurement of the longitudinal modulus for each specimen is unnecessary. The influence of specimen elastic modulus variability on the measured toughness is taken into account, which is advantageous. During propagation, an equivalent crack length (a_e) can be used to include the effects of a non-negligible FPZ. This requires the solution of the cubic equation (5.1) following the methodology presented in Section 3.1.4. Toughness under mixed-mode I + II loading can be obtained combining the Irwin–Kies relation (Eq. 2.39) with Eq. (5.1)

$$G_T = \frac{21P^2 a_e^2}{16E_f B^2 h^3} + \frac{3P^2}{40G_{LR}B^2 h} \tag{5.3}$$

According to the partition method proposed by Szekrényes and Uj (2004), the mode I and II components become

$$G_I = \frac{12P^2 a_e^2}{16E_f B^2 h^3} + \frac{3P^2}{40G_{LR}B^2 h}; \quad G_{II} = \frac{9P^2 a_e^2}{16E_f B^2 h^3} \tag{5.4}$$

Using this procedure, the total energy and its components do not depend on crack length measurement during propagation. They only depend on the specimen compliance, which is used to estimate an equivalent crack, hence accounting for the FPZ effects. In fact, the FPZ influences the profile of the load–displacement curve, thus affecting the specimen compliance.

5.1.3 Numerical Analysis

With the purpose of verifying the performance of the proposed data reduction scheme, a numerical analysis was performed. The numerical model consists of a plane stress analysis using 3,840 eight-node isoparametric solid elements and six-node 413 cohesive elements. The cohesive damage model is based on linear softening relationship between stresses and relative displacements under mixed-mode loading (Figure 4.7). The pure mode laws (I and II) were defined from average values of toughness and local strengths determined in previous chapters ($\sigma_{1,I} = 5.34\,\text{MPa}$, $G_{Ic} = 0.264\,\text{N/mm}$, $\sigma_{1,II} = 9.27\,\text{MPa}$ and $G_{IIc} = 0.94\,\text{N/mm}$). The linear energetic criterion (Eq. 2.50) was used to deal with the mixed-mode loading (Figure 5.3).

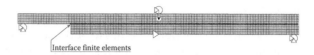

FIGURE 5.3
FE mesh used for the simulation of the SLB test.

The values of applied load, displacement and crack length were registered for each increment during propagation. Two different data reduction methods were used to obtain the *Resistance*-curves (*R*-curves): the CBBM and the compliance calibration method (CCM). Numerically, the CCM can be straightforwardly applied, since the crack length during its growth is easily accessed by the coordinate of the first 'not opened' point at the crack tip. The compliance calibration as a function of crack length a was performed considering a cubic polynomial $C = A_3 a^3 + A_2 a^2 + A_1 a + A_0$. In the following, the application of the Irwin–Kies relation (Eq. 2.39) becomes simple; basically, it only depends on the differentiation of the adjusted polynomial. Figure 5.4 shows the numerical *R*-curves obtained for the SLB specimen. Excellent agreement between both methods for $G_T = G_I + G_{II}$ in the

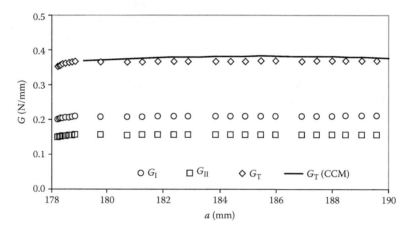

FIGURE 5.4
R-curves provided by the CBBM and CCM (Oliveira et al., 2009).

TABLE 5.1

Summary of the SLB Experimental Results

a_0 (mm)		G_{Tc} (N/mm)	G_I (N/mm)	G_{II} (N/mm)	G_I/G_{II}
168	Average	0.385	0.221	0.165	1.34
	CoV (%)	17.2			
183	Average	0.390	0.223	0.167	1.34
	CoV (%)	13.2			
198	Average	0.388	0.222	0.166	1.34
	CoV (%)	17.8			
213	Average	0.396	0.226	0.169	1.34
	CoV (%)	13.6			
	Global avg.	0.390	0.223	0.167	1.34

Mixed-Mode I + II Fracture Characterisation

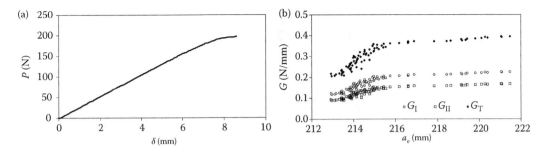

FIGURE 5.5
(a) Typical P–δ curve and (b) the corresponding R-curves of the SLB test.

plateau region was verified. The obtained values for G_{Tc}, 0.37 N/mm using the CBBM and 0.38 N/mm using the CCM, are in close agreement with the average experimental value (see Table 5.1). The same conclusion can be drawn for mode components using the CBBM, $G_I = 0.21$ N/mm and $G_{II} = 0.16$ N/mm.

5.1.4 Experimental Results

The experimental load–displacement (P–δ) curves were recorded for each tested specimen (Figure 5.5a). The respective R-curves for each mode component and for total G (G_T) were obtained using the CBBM (Figure 5.5b). The value of fracture energy under mixed-mode ($G_{Tc} = G_I + G_{II}$) was taken from the plateau of the G_T R-curve (Eq. 5.3) and characterises the self-similar propagation process. The respective mode I and II components were obtained using Eq. (5.4).

Table 5.1 summarises the consistent results obtained considering at least 10 tests for each case. In fact, a coefficient of variation (CoV) inferior to 20% was observed, which is good in wood. It can be seen that toughness does not depend on the initial crack length. In addition, a constant mode ratio of 1.34 was obtained. With the purpose of varying the mode ratio, specimens with different arm thickness should be considered. However, a limited range would be achieved since thin arms fail under bending. It can be concluded that SLB test provides fracture under mixed-mode I+II loading for an almost constant mode ratio, which can be considered a limitation of this test.

5.2 End Load Shear-Mixed Mode Test

5.2.1 Test Description

The ELS-MM is similar to ELS test used for fracture characterisation under pure mode II loading. It consists of a cantilever beam with a clamping fixture system that slides horizontally to avoid stretching loads on the specimen. The inexistence of the superior specimen arm (Figure 5.6) induces a mixed-mode I + II loading at the crack tip.

With the purpose of analysing the influence of the initial crack length on toughness, four values of a_0 were used in this study (Figure 5.7). The habitual procedure was followed to induce the pre-crack and perform its measurement.

FIGURE 5.6
Experimental setup for the ELS-MM test (Oliveira et al., 2009).

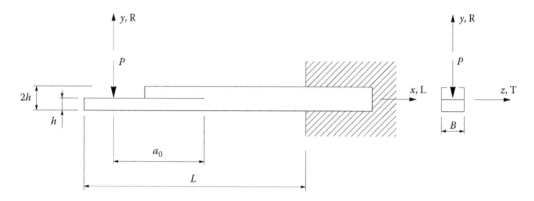

FIGURE 5.7
Schematic representation of the ELS-MM test ($2h = 20$, $L = 175$, $B = 20$ and $a_0 = 60, 90, 120$ and 150; all dimensions in mm).

5.2.2 Compliance-Based Beam Method

As in previous fracture tests, an equivalent crack length procedure based on specimen compliance and beam theory was followed. Using the Timoshenko beam theory, the specimen compliance writes (Oliveira et al., 2009)

$$C = \frac{7a^3 + L^3}{2E_L B h^3} + \frac{3(a+L)}{5G_{LR} B h} \tag{5.5}$$

The application of the CBBM in the ELS-MM test is subjected to an extra difficulty associated to imperfect real clamping conditions. As a result, an effective beam length (L_{ef}) satisfying the conditions of a perfect clamp must be defined in Eq. (5.5) instead of the real L. The known initial conditions (a_0 and C_0) can be utilised to write

$$\frac{L_{ef}^3}{2Bh^3 E_L} + \frac{3L_{ef}}{5BhG_{LR}} = C_0 - \frac{7a_0^3}{2Bh^3 E_L} - \frac{3a_0}{5BhG_{LR}} \tag{5.6}$$

allowing to eliminate L_{ef} in Eq. (5.5)

$$C = C_0 + \frac{7(a^3 - a_0^3)}{2E_L B h^3} + \frac{3(a - a_0)}{5G_{LR} B h} \tag{5.7}$$

In the course of the test, the equivalent crack length can be estimated from previous equation through the formulae presented in Section 3.1.4. Finally, the evolution of the total strain energy release rate can be easily obtained combining the Irwin–Kies relation (Eq. 2.39) with Eq. (5.7)

$$G_T^{\text{ELS-MM}} = \frac{21 P^2 a_e^2}{4 E_L B^2 h^3} + \frac{3 P^2}{10 G_{LR} B^2 h} \tag{5.8}$$

Using the partition modes procedure proposed by Szekrényes and Uj (2004), the modes I and II components come to be

$$G_I^{\text{ELS-MM}} = \frac{12 P^2 a_e^2}{4 E_L B^2 h^3} + \frac{3 P^2}{10 G_{LR} B^2 h}; \quad G_{II}^{\text{ELS-MM}} = \frac{9 P^2 a_e^2}{4 E_L B^2 h^3} \tag{5.9}$$

The ELS-MM test presents a disadvantage relative to SLB associated with the obligatory measurement of specimen longitudinal modulus (E_L) in a three-point bending test, before performing the fracture test. In fact, the inherent variability of elastic properties between specimens cannot be neglected since it markedly affects the measured values. Other aspects intrinsic to equivalent crack procedure, as is the case of accounting for a non-negligible FPZ, are considered in the CBBM of the ELS-MM test.

5.2.3 Numerical Analysis

A numerical analysis with finite element model was performed. Specimen arms were simulated considering 3,840 plane eight-node isoparametric solid elements, and the loading displacement was applied through a rigid sphere. Perfect clamping conditions were considered at the specimen's right extremity considering the effective specimen length (L_{ef}). The numerical validation procedure involved 348 six-node cohesive elements located at the specimen mid-plane. As in the previous case (see SLB test), the linear softening cohesive law (Figure 4.7) and the linear energetic criterion (Eq. 2.50) were used to deal with the mixed-mode loading. Also, the pure mode cohesive parameters defined in previous chapters ($\sigma_{u,I} = 5.34$ MPa, $G_{Ic} = 0.264$ N/mm, $\sigma_{u,II} = 9.27$ MPa and $G_{IIc} = 0.94$ N/mm) were considered in this analysis (Figure 5.8).

The CCM and CBBM were both applied to numerical results ensuing from simulations. From Figure 5.9, it can be concluded that both methods provide consistent results. The values for G_{Tc} point to 0.363 N/mm using the CBBM and 0.375 N/mm using the CCM, which are lower than the average experimentally measured values (see Table 5.2). This can be explained by some experimental problems inherent to this test, like large displacement and rotation effects, that can affect the accuracy of beam theory based methods.

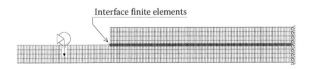

FIGURE 5.8
FE mesh used for the simulation of the ELS-MM test.

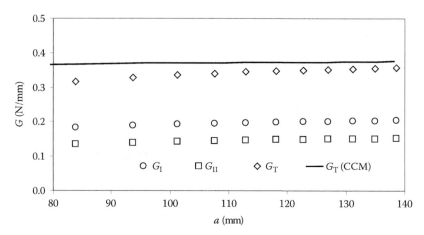

FIGURE 5.9
R-curves obtained numerically for the ELS-MM specimen by the CBBM (G_I, G_{II} and G_T) and CCM [G_T (CCM)] (Oliveira et al., 2009).

TABLE 5.2

Summary of the ELS-MM Experimental Results

a_0 (mm)		G_{Tc} (N/mm)	G_I (N/mm)	G_{II} (N/mm)	G_I/G_{II}
60	Average	0.404	0.233	0.171	1.37
	CoV (%)	15.0			
90	Average	0.417	0.240	0.177	1.35
	CoV (%)	11.4			
120	Average	0.411	0.236	0.175	1.34
	CoV (%)	14.8			
150	Average	0.391	0.224	0.167	1.34
	CoV (%)	18.3			
	Global avg.	0.406	0.233	0.173	1.35

5.2.4 Experimental Results

Ten tests were performed for each initial crack length, leading to a total of 40 ELS-MM tests. The experimental load–displacement (P–δ) data (Figure 5.10a) were used in CBBM to get the corresponding R-curves (Figure 5.10b) of mode I and II components (Eq. 5.9), as well as the one corresponding to the total strain energy release rate (G_T) under mixed mode (Eq. 5.8).

The global results are condensed in Table 5.2. Once again, the CoV is inferior to 20%, revealing the soundness of the experimental results. The small variation of mode ratio is not relevant and probably dictated by slight testing spurious variability.

Several conclusions can be drawn from the comparison between the results of the SLB and ELS-MM tests. The SLB test is easier to be executed than the ELS-MM, since it does not require an independent measurement of longitudinal modulus and the experimental setup is also simpler. The CoV of toughness for each initial crack length analysed is lower than 20%, which is good in wood. There is no clear initial crack length dependency of the measured fracture energy. In fact, in both tests, the average values obtained for each initial crack length are approximately constant. The mode ratio is also approximately constant.

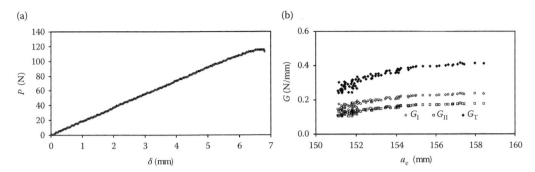

FIGURE 5.10
(a) Typical $P-\delta$ curve and (b) the corresponding R-curve of the ELS-MM test (Oliveira et al., 2009).

In effect, although some mode ratio variation could be achieved considering specimens with different arm thickness, this would be confined to a quite limited range, since thin specimen arms fail under bending.

Finally, the global average values provided by both tests agree with each other. Owing to these similarities, it was decided to use the global average value of toughness provided by the SLB test in a G_I versus G_{II} space considering the critical pure mode loading values ($G_{Ic} = 0.264$ N/mm and $G_{IIc} = 0.94$ N/mm) determined in previous chapters. From Figure 5.11, it can be concluded that the linear energetic criterion (Eq. 2.50) fits the observed trend well, thus constituting an indication that this criterion is adequate to represent the fracture envelop of *Pinus pinaster* Ait. wood species.

5.3 Mixed-Mode Bending Test

5.3.1 Test Description

The MMB test was developed by Reeder and Crews (1990) in the context of mixed-mode I + II interlaminar fracture characterisation of artificial composites. The MMB test can be viewed as a combination of DCB and ENF tests. In effect, it consists of a three-point bending test, as in the case of ENF, with an opening load displacement at the specimen

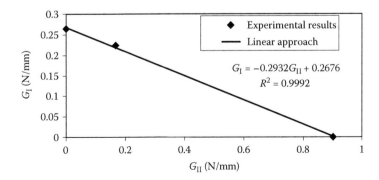

FIGURE 5.11
Representation of fracture energy of *Pinus pinaster* Ait. wood in G_I versus G_{II} space (Oliveira et al., 2009).

extremity containing the pre-crack (Figures 5.12a and b). Henceforth, the load (*P*) applied at the end of the loading lever (Figure 5.12b) leads to a downward load at the specimen mid-span, inducing the mode II loading and an upward applied load at the specimen cracked extremity producing mode I.

The loading ensuing from static equilibrium of the specimen can be separated into mode I and mode II components (Figure 5.13), taking into account that MMB test is a combination of DCB and ENF tests

$$P_I = \left(\frac{3c-L}{4L}\right)P; \quad P_{II} = \left(\frac{c+L}{L}\right)P \quad (5.10)$$

The main advantage of the MMB test is the possibility of altering the mode ratio in a relatively wide range in the $G_I - G_{II}$ space, thus providing a simple way to obtain a detailed fracture envelop under mixed-mode I+II loading; it should be emphasised that this is exactly the major drawback of the previous described SLB and ELS-MM tests. The changing of the mode ratio is achieved by varying the length (parameter *c* in Figures 5.12b and 5.13) of the loading lever. In reality, from Eq. (5.10), it can be observed that both load components (P_I and P_{II}) depend directly on the value of *c*.

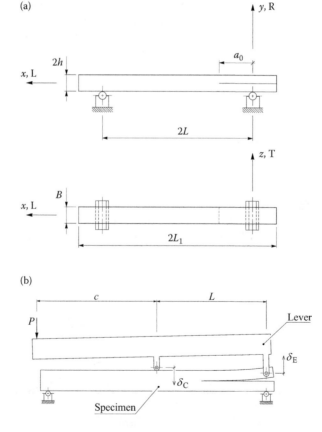

FIGURE 5.12
(a) Specimen dimensions in mm ($2L_1 = 500$, $2L = 460$, $2h = 20$, $B = 20$, $a_0 = 162$) and (b) schematic representation of the MMB test.

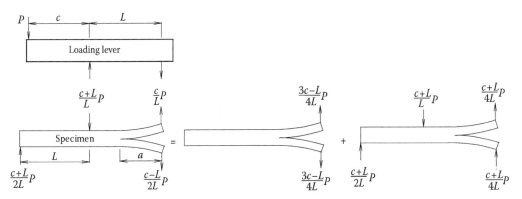

FIGURE 5.13
Static equilibrium and load mode partition of the MMB specimen (de Moura et al., 2010).

This partition of loads (P_I and P_{II} as a function of P) makes available a straightforward application of the already developed crack equivalent based methods (CBBM) for DCB and ENF tests, to get the R-curves corresponding to mode I and mode II components. With this aim, the displacement at the specimen extremity ($\delta_E = \delta_I$) should be measured during the test to define $C_I = \delta_I/P_I$. Likewise, the $C_{II} = \delta_{II}/P_{II}$ is defined from $\delta_{II} = \delta_C + \delta_I/4$ (Reeder and Crews, 1990), which requires the monitoring of the mid-span displacement δ_C. From C_I and C_{II}, the CBBM can be easily applied considering Eqs. (3.8–3.14) and Eqs. (4.8–4.11) for mode I and mode II strain energy release rate components, respectively (Oliveira et al., 2007). Hence, for each mode ratio, the values of energy components during self-similar crack growth are given by the plateaus of the corresponding R-curves. This procedure simplifies drastically the application of the MMB test to wood, since no crack monitoring is necessary and FPZ effects are taken into account (de Moura et al., 2010).

5.3.2 Experimental Analysis and Results

A special device considering larger dimensions relative to the dispositive initially developed for interlaminar fracture characterisation of composites (Reeder and Crews, 1990) was conceived (Figure 5.14). The loading lever is made using an aluminium alloy, thus minimising the spurious effects due to its weight.

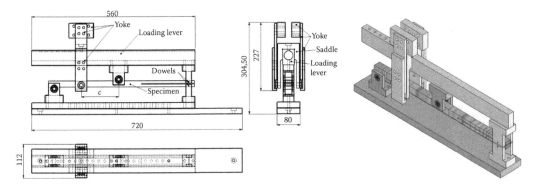

FIGURE 5.14
Design of the MMB test setup for wood testing. (Adapted from de Moura et al., 2010.)

Two LVDTs (Figure 5.15) were used to measure the displacements at the specimen edge (δ_E) and at the mid-span (δ_C), leading to δ_I and δ_{II} necessary to apply the CBBM. The load was applied with a crosshead speed of 5 mm/min, being transmitted to the specimen via bearing-mounted roller at the specimen mid-span, and a dowel connection to induce the crack opening at the specimen extremity.

Ten different mode ratios (G_I/G_{II}) were analysed: 0.05, 0.1, 0.15, 0.25, 0.5, 0.75, 1.0, 1.25, 2.0 and 2.75, which is the maximum possible value provided by the designed setup owing to large specimen dimensions used in wood. Ten tests were performed for each mode ratio, and at least seven valid results were obtained.

Figure 5.16a plots typical load–displacement curves considering the loading mode partition discussed earlier. These curves were used in the CBBM to achieve the corresponding R-curves of both mode components (Figure 5.16b). The plateau values provide an easy determination of mode ratio (G_I/G_{II}) as well as the critical fracture energy for this mixed-mode loading ($G_{Tc} = G_I + G_{II}$) under self-similar crack growth.

Table 5.3 summarises all the toughness values resulting from this MMB testing campaign. The CoV was, in general, larger than in previous fracture tests. The complex

FIGURE 5.15
Experimental setup for the MMB test applied to wood (de Moura et al., 2010).

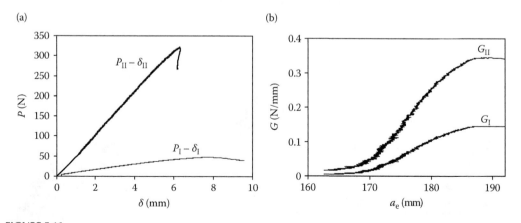

FIGURE 5.16
(a) Typical P–δ curves obtained considering a mode ratio of 0.5 and (b) the corresponding R-curves (de Moura et al., 2010).

TABLE 5.3

Summary of the MMB Experimental Results

G_I (N/mm)	G_{II} (N/mm)	G_{Tc} (N/mm)	G_{Tc} CoV (%)	G_I/G_{II}
0.036	0.723	0.759	26.2	0.05
0.056	0.633	0.690	12.7	0.09
0.073	0.544	0.616	28.5	0.13
0.124	0.502	0.626	27.3	0.25
0.161	0.356	0.517	14.5	0.46
0.196	0.278	0.474	27.4	0.71
0.223	0.220	0.443	27.0	1.01
0.257	0.212	0.469	27.4	1.21
0.254	0.135	0.389	23.8	1.89
0.258	0.101	0.359	23.6	2.65

Source: Adapted from de Moura et al. (2010).

micro-mechanisms of rupture involved during a mixed-mode fracture, which vary as a function of mode ratio, is a possible explanation.

The 10 mixed-mode ratios were plotted in the $G_I - G_{II}$ space together with the average pure mode critical values ($G_{Ic} = 0.264$ N/mm and $G_{IIc} = 0.94$ N/mm). Three different criteria were utilised to represent the observed trend: the linear energetic (Eq. 2.50), the quadratic one (Eq. 2.43 with $\alpha = \beta = 2$) and the Benzeggagh–Kenane (1996) criterion, usually known as the B–K criterion

$$G_{Tc} = G_{Ic} + (G_{IIc} - G_{Ic}) + \left(\frac{G_{II}}{G_T}\right)^\eta \quad (5.11)$$

with $\eta = 1.5$, which was verified to be the best fitting value. Figure 5.17 plots the experimental values and the lines corresponding to the referred three fracture criteria.

In the region of predominant mode I, the quadratic and B–K criteria present good agreement with the experimental values, and the linear criterion is somewhat conservative. In fact, the quadratic and B–K criteria present a plateau in this region of $G_I - G_{II}$ space, revealing that a small presence of mode II practically does not alter the mode I component,

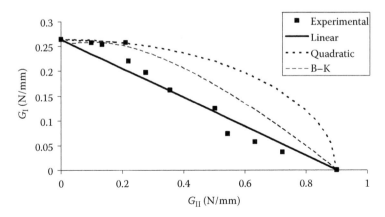

FIGURE 5.17
Representation of fracture energy of *Pinus pinaster* Ait. wood in $G_I - G_{II}$ space (de Moura et al., 2010).

which is not the case of the linear criterion that underestimates the fracture energy. For the remaining cases, the linear criterion presents the best performance and the other ones become non-conservative. In general, it can be concluded that the linear criterion provides the best agreement with the experimental values, thus being selected to be used as representative of the fracture envelop of the *Pinus pinaster* Ait. wood species.

5.3.3 Numerical Validation

The application of the MMB test with the equivalent crack length based reduction schemes is a novel procedure concerning wood fracture characterisation under mixed-mode I + II loading. In this context, a numerical validation using plane stress finite element analysis (Figure 5.18) with 1,312 eight-node isoparametric solid elements and 164 six-node cohesive elements placed at the specimen mid-plane was carried out. Very small increments and refined mesh in the region of crack propagation contributed to a smooth crack growth. Rigid beam-type connections were assumed between the loading, central and extremity pins.

The validation was performed just for one mode ratio. Since the linear energetic criterion was implemented in the model, the $G_I/G_{II} = 0.25$ was chosen as it reveals good agreement with this criterion (Figure 5.17). Figure 5.19 shows an excellent agreement between the numerical R-curves and the experimental average values of this mode ratio. This means that all the proposed procedure is valid when applied to the MMB test concerning wood fracture characterisation under mixed-mode I + II loading.

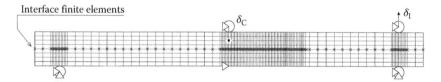

FIGURE 5.18
FE mesh used for the simulation of the MMB test.

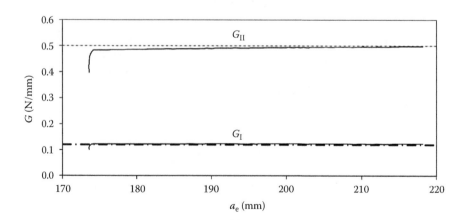

FIGURE 5.19
Comparison between the numerical R-curves (solid lines) and the average experimental values of strain energy (dashed lines) for $G_I/G_{II} = 0.25$ (de Moura et al., 2010).

5.4 Conclusions of Mixed-Mode I + II Fracture Tests

Three fracture tests were applied in the context of mixed-mode I + II fracture characterisation of wood. Equivalent crack length based data reduction schemes were developed for all the tests, thus making them more accurate and straightforward to apply. The SLB and ELS-MM tests are very simple to execute, but they present the disadvantage of providing an almost fixed mode ratio. This aspect renders difficult the definition of a global fracture envelop in the $G_I - G_{II}$ plane. In addition, the ELS-MM requires a special testing device and the pre-measurement of the elastic modulus. The MMB test permits accomplishing tests under a wide range of mode ratios only varying the lever arm length. This aspect is fundamental for a rigorous fracture envelop determination of a material under mixed-mode I + II loading, which makes the MMB test the best option, even if the experimental apparatus is somewhat complicated and with large dimensions.

Finally, it must be emphasised that all the tests point to the linear energetic criterion as being appropriate to simulate the fracture envelop of the *Pinus pinaster* Ait. wood species under mixed-mode I + II loading.

References

Benzeggagh, M. L. and M. Kenane (1996). Measurement of mixed-mode delamination fracture toughness of unidirectional glass/epoxy composites with mixed-mode bending apparatus. *Compos Sci Technol*, 56:439–49.

de Moura, M. F. S. F., J. M. Q. Oliveira, J. J. L. Morais and J. M. C. Xavier (2010). Mixed-mode I/II wood fracture characterization using the mixed-mode bending test. *Eng Fract Mech*, 77:144–52.

Jernkvist, L. O. (2001). Fracture of wood under mixed-mode loading II. Experimental investigation of *Pices abies*. *Eng Fract Mech*, 68:565–76.

Oliveira, J. M. Q., M. F. S. F. de Moura, J. J. L. Morais and M. A. L. Silva (2007). Numerical analysis of the MMB test for mixed-mode I/II wood fracture. *Compos Sci Technol*, 67:1764–71.

Oliveira, J. M. Q., M. F. S. F. de Moura and J. J. L. Morais (2009). Application of the end loaded split and single-leg bending tests to the mixed-mode fracture characterization of wood. *Holzforschung*, 63:597–602.

Reeder, J. R. and J. H. Crews (1990). Mixed-mode bending method for delamination testing. *AIAA J*, 28:1270–6.

Szekrényes, A. and J. Uj (2004). Beam and finite element analysis of quasi-unidirectional composite SLB and ELS specimens. *Compos Sci Technol*, 64:2393–406.

Tschegg, E. K., A. Reiterer, T. Pleschbergers and E. Stanzl-Tschegg (2001). Mixed-mode fracture energy of spruce wood. *J Mater Sci*, 36:3531–7.

6

Structural Applications – Case Studies

Wood and wood products are among the most important construction materials. Wood is generally used in frames, buildings, truss roof structures in buildings, bridges, towers, railroad infrastructures and many more applications. Damage and failure behaviour of wood members in tensile, compressive or shear loading are extremely important to account for in wooden structures subjected to high working stresses. Structural details involving wood member's connections also require special attention for a safe design. For example, damage under tensile, compressive or shear loading can occur at the joints (connector plates) or within the lumber members. Consequently, predictive methods and models for the simulation of the structural behaviour of these elements, such as finite element analysis with cohesive zone modelling, can be viewed as being essential tools.

The objective of this chapter is to describe some structural applications involving the concepts detailed in previous chapters. The selected cases focussed on structural details like wood connections considering repair/reinforcement strategies using bonded carbon-epoxy laminates and dowel joints. All examples involve numerical analysis with cohesive zone modelling and experimental testing for the sake of model validation.

6.1 Wood Bonded Joints

Owing to its biological origin, wood members are frequently affected by the presence of knots, checks, shakes, splits, slope of grain, reaction wood and decay and other defects. Premature replacement of structural members should be avoided since it can be expensive and difficult to perform. Hence, repairing or reinforcing wood structural members with artificial composites can be a valuable alternative. In the following, some examples of such strategies and the utility of cohesive zone modelling on their optimisation are described.

6.1.1 Repaired Beam under Tensile Loading

In this case, a *Pinus pinaster* wood beam under tension is supposed to have suffered any kind of damage, such as tensile failure induced by overloads, natural decay or human intervention. The repair involves the replacement of the portion of damaged wood by an insert of the same material (Barreto et al., 2010) adhesively bonded to the sane wood parts. Additionally, the joint is reinforced with bonded unidirectional carbon-epoxy composite patches (four plies) (Figure 6.1) using Araldite 2015®. The bonding technique is identified as the most efficient method of stress transfer between artificial composites and wood, as it avoids the stress concentrations intrinsic to connection with mechanical fasteners. The axes (x, y, z) of Figure 6.1 correspond to the fibre, thickness and transverse directions of the carbon-epoxy patch, respectively. The (L, R, T) coordinate system identifies wood orientations.

105

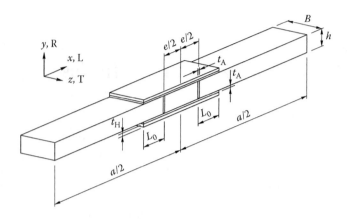

FIGURE 6.1
Schematic representation of the repair with the following dimensions in millimetres: $B = 20$, $h = 10$, $a = 200$, $e = 20$, $t_A = 0.2$, $t_H = 0.6$ and $L_o = (5, 10, 15)$. (Adapted from Barreto et al., 2010.)

Three different overlap lengths ($L_o = 5, 10, 15\,\text{mm}$) were considered in this study (Barreto et al., 2010). Specimens were tested (Figure 6.2) under displacement control with a displacement rate of 1 mm/min. Figure 6.3 reveals the typical failure mode observed independently of the overlap length used. Typically, fracture occurred simultaneously in the wood beam close to the wood/adhesive interface in the overlap region and vertically, between the wood member and the insert as a cohesive failure of the adhesive layer. The horizontal fracture in wood was initiated by a vertical crack (LR plane) close to the patch edge that subsequently propagated horizontally.

In the numerical analysis, several failure paths with different cohesive laws were considered on the made-up mesh of finite elements. A schematic representation of the numerical model is presented in Figure 6.4, revealing the location of cohesive elements

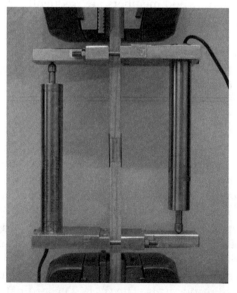

FIGURE 6.2
Experimental setup for $L_o = 5\,\text{mm}$ case (Barreto et al., 2010).

Structural Applications – Case Studies

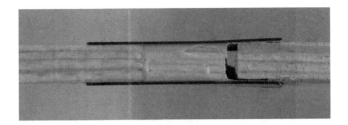

FIGURE 6.3
Example of experimental fractures ($L_o = 15\,mm$). (Adapted from Barreto et al., 2010.)

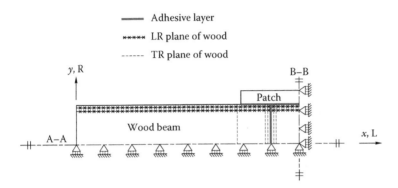

FIGURE 6.4
Position of the cohesive elements with different cohesive laws. (Adapted from Barreto et al., 2010.)

with different properties (Table 6.1) to mimic diverse failure modes. The trapezoidal with linear softening law for mixed-mode I + II loading (Figure 2.16) was considered to simulate fracture inside the Araldite layer. In the case of wood, the linear cohesive law (Figure 4.7) was adopted. The mesh was built to allow alteration of the failure path, thus including several different possibilities of collapse. Symmetry conditions were imposed to simplify the model and reduce the CPU time consumption. Plane-stress eight-node isoparametric rectangular solid elements were used to model the wood beam and the patch and six-node cohesive elements were placed at the aforementioned interfaces.

Figure 6.5 shows the failure modes obtained in the numerical analysis. As observed experimentally, failure initiates by a vertical crack in the wood external layer occurring in the vicinity of the patch edge followed by horizontal failure inside wood. This can be explained by stress concentrations that develop at this singularity region and the lower

TABLE 6.1
Cohesive Parameters in Pure Modes (I, II) Used to Simulate Different Failures (see Figure 4.7)

Cohesive Laws	Mode i	G_{ic} (N/mm)	$\sigma_{1,i}$ (MPa)	$\delta_{2,i}$ (mm)	$\delta_{u,i}$ (mm)
Adhesive layer (Araldite® 2015)	I	0.43	23.0	0.0187	0.021
	II	4.7	22.8	0.1710	0.248
Wood RL fracture system	I	0.2	16.0	1.6×10^{-5}	0.025
	II	1.2	16.0	1.6×10^{-5}	0.150
Wood LR fracture system	I	25	65.0	6.5×10^{-5}	0.77
	II	1.2	16.0	1.6×10^{-5}	0.15

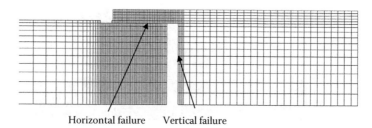

FIGURE 6.5
Detail of the failure mode obtained numerically (Barreto et al., 2010).

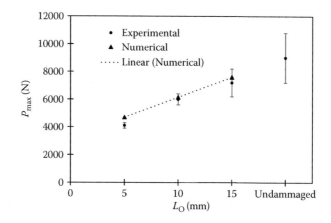

FIGURE 6.6
Evolution of maximum load (P_m) as a function of overlap length L_o.

interlaminar shear strength of wood relative to adhesive. The vertical cohesive failure at the adhesive layer between the wood member and the insert is justified by the smaller peel strength of the adhesive relative to wood strength in the L-direction. This circumstance is a symptom that undamaged strength of the beams, which correspond roughly to a cross-sectional failure outside the repaired region, was not achieved for the values of overlap length considered.

Figure 6.6 plots the evolution of maximum load achieved in repaired beams as a function of the overlap length. It was verified as a rising trend with the increase of overlap length, but the maximum value considered ($L_o = 15$ mm) is insufficient to replicate the strength of undamaged beams.

The wood and carbon-epoxy patches were modelled as elastic orthotropic materials, whose elastic properties are presented in Table 6.2.

TABLE 6.2

Elastic Properties of Pine Wood and Carbon-Epoxy (C-E) Laminate

Wood	E_L (GPa)	$E_R = E_T$ (GPa)	$\nu_{LR} = \nu_{LT}$	ν_{RT}	G_{LR} (GPa)	G_{LT} (GPa)	G_{RT} (GPa)
	10.2	1.01	0.342	0.38	1.1	1.1	0.17
C-E	E_1 (GPa)	$E_2 = E_3$ (GPa)	$\nu_{12} = \nu_{13}$	ν_{23}	G_{12} (GPa)	G_{13} (GPa)	G_{23} (GPa)
	109	8.82	0.342	0.38	4.32	4.32	3.2

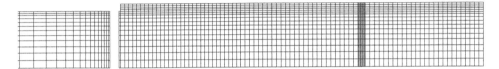

FIGURE 6.7
Failure mode predicted numerically for a repair with $L_o = 20.80$ mm (Barreto et al., 2010).

Anyway, it can be concluded that numerical results are in close agreement with the experimental ones, which validates the numerical procedure. Taking this aspect into account it was decided to determine numerically the value of L_o leading to the average maximum load obtained experimentally (around 9,000 N) and simultaneously altering the failure mode to beam transverse cross-sectional failure near the patch edge (Figure 6.7). It was verified that for $L_o = 20.8$ mm such conditions are satisfied, meaning that a repair efficiency of 100% is possible, i.e. full strength recovery is attained for this structural component repaired with two patches of carbon-epoxy composite.

6.1.2 Repaired Beam under Bending Loading

The objective of this study is to analyse the effect of carbon-epoxy composite patch bonding repairs in wood beams damaged by cross-grain failure and submitted to bending (Dourado et al., 2012). Because of grain misalignment, this type of damage develops inclined relatively to wood beam axis frequently (Figure 6.8). This circumstance leads to a significant reduction of beam flexural strength, especially when damage is located on beam tensile region. Repair of these wood damaged beams under bending using carbon-epoxy composites has been analysed. The influence of patch thickness and adhesive filleting was

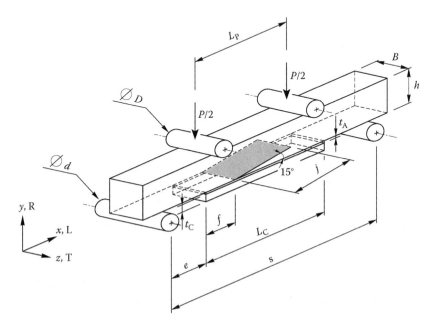

FIGURE 6.8
Repaired beam geometry under four-point bending. Wood anatomic axis: L = Longitudinal; R = Radial; T = Tangential ($d = 15$, $D = 60$, $e = 100$, $f = 21$, $L_C = 62$, $j = 20$, $s = 20$ mm). (Adapted from Dourado et al., 2012.)

evaluated experimentally and numerically. The main goal was to assess the effect of these parameters on the ultimate load and failure type.

Pinus pinaster Ait. was used in this work as the testing material. A total of 25 specimens were machined from a single wood log and stabilised at laboratory conditions before the experiments. The selected pre-crack angle (15°) intends to simulate severe grain misalignment in wood beams, which defines natural paths prone to crack propagation, and it was executed using a circular band saw (1mm thick). Unidirectional composite laminates were manufactured from high strength carbon pre-preg (Texipreg HS 160 RM from SEAL®) following the manufacturer recommendations using a hot-plate press. Patches were cut from laminated plates with two different thicknesses (0.6 and 2.0mm) and bonded onto beam-damaged region using SIKADUR 30 structural adhesive from Sika®. This adhesive presents a glass transition temperature of 62°C, which is adequate for typical timber applications. Surfaces were duly cleaned and sandpapered (180-grit) to improve bonding quality, avoiding spurious adhesive failure at the interfaces (adhesive/composite and adhesive/wood).

The four-point-bending tests were performed using a mechanical spindle-driven tension-compression machine (Instron 1125) under displacement control and with crosshead displacement rate of 0.3 mm/min. A two-point assembly device with rotating cylindrical contact surfaces was used in the mechanical tests, inducing equal reactions in the beam supports. Figure 6.9 reveals the typical failure mode observed in an unrepaired beam, which is characterised by the development of an extensive longitudinal crack, leading to a significant decrease of the bearing load.

Subsequently, repaired specimens using a 2mm thick patch bonded to the beam tensile region were tested. Typically, the characteristic failure starts at the corner patch left side (Figure 6.10) and propagates through wood near the wood-adhesive interface.

Failure path inside wood is explained by high peel stresses that develop at the geometrical singularity. These stresses are aggravated by the mismatch stiffness between wood and carbon-epoxy composite. Hence, to diminish this stress concentration effect and contribute to material saving, a thinner patch ($t_C = 0.6$mm which corresponds to a laminate with four layers of carbon-epoxy) leading to an average increase of 32% on the initial stiffness was considered in a second series of tests. A stress analysis performed by means of finite element method revealed a reduction of 15% in the peel stress at this critical

FIGURE 6.9
Failure path of an unrepaired beam (Dourado et al., 2012).

Structural Applications – Case Studies

FIGURE 6.10
Failure path of repaired beam with thick patch (Dourado et al., 2012).

point and showed that developed longitudinal stresses in the patch are below the laminate strength. As a result of the stress concentration reduction, the crack did not propagate inside wood but rather along the adhesive (Figure 6.11), which is a consequence of the peel stress reduction attained when a thinner patch is used.

To induce further reduction of stress concentration effects, an adhesive fillet of 30° was chosen in a new series of tests for the considered two patch thicknesses (i.e. t_C = 2.0 and 0.6 mm). This fillet was made by means of a specially fabricated mould to guarantee the inclination of 30°, shaping the adhesive in contact with wood. Like previous cases (i.e. repairing without fillet), the crack propagated in the adhesive (thin patch) and inside wood (thick patch). Nevertheless, in both circumstances, the failure initiated under mode I loading by debonding in the vertical boundary between the patch and the fillet (Figure 6.12).

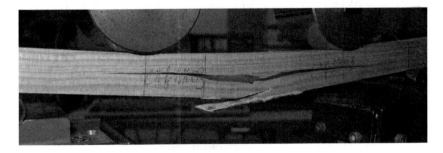

FIGURE 6.11
Failure path of repaired beam with thin patch (Dourado et al., 2012).

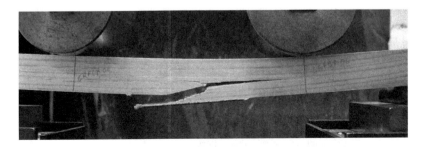

FIGURE 6.12
Failure path of repaired beam with thin patch with fillet (Dourado et al., 2012).

The average values of the ultimate load obtained for all tested cases are summarised in Figure 6.13. It can be observed that the repair increases the damaged beam bearing load pronouncedly. Despite of the observed scatter in the P–δ curves, it was verified that thin patches are slightly more efficient than thicker ones. Finally, it can be concluded that the use of adhesive fillets does not lead to a significant increase in the failure load. The explanation is related to the way as crack initiates; as already referred, in the vertical boundary between the patch and adhesive fillet. This boundary region is affected by the important mismatch stiffness between the patch and the adhesive, and it is also under remarkable tensile stresses due to bending. Therefore, premature failure occurs in this boundary, precluding an evident gain on the bearing load by adding the adhesive fillet.

A two-dimensional numerical analysis considering mixed-mode I + II cohesive zone modelling was also performed considering the linear cohesive law (Figure 4.7). The model accounts for several possible crack paths (Figure 6.14) to assess its ability to capture the real failure mode. Different cohesive properties were used to mimic failure inside wood in the RL and LR plane, inside the patch (carbon-epoxy) and in the adhesive

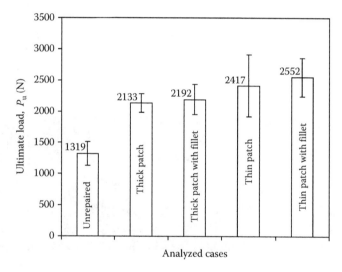

FIGURE 6.13
Average values of ultimate load considering five specimens for each case (Dourado et al., 2012).

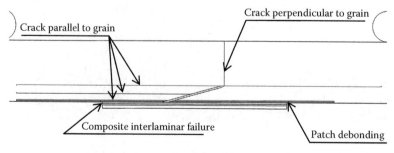

FIGURE 6.14
Crack paths considered in finite element analysis (Dourado et al., 2012).

(see Tables 6.1 and 6.3). In addition, several different paths were also considered in the fillet region (Figure 6.15).

Figure 6.16 illustrates the crack paths obtained numerically for repaired beams with thick patch, thin patch and thin patch with fillet. Comparing the failure modes obtained numerically with the experimental ones (Figures 6.10–6.12), it can be concluded that the model behaved well in this aspect. In addition, a quantitative analysis regarding the initial stiffness, maximum load attained (P_u) and post-peak behaviour demonstrated the ability of the proposed finite element analysis including cohesive zone modelling to reproduce the real behaviour observed (Figure 6.17).

In conclusion, it can be affirmed that repairing wood beams damaged by cross-grain failure under bending using carbon-epoxy composites is quite effective. The best solution (thin patch with fillet) leads to an almost double failure load relative to damaged and unrepaired beam.

6.1.3 Reinforcement of Wood Structures

Reinforcement of wood beams with artificial composites to increase their stiffness and strength is frequently used in structural applications (Alhayek and Svecova, 2012; Borri et al., 2005; Motlagh et al., 2008; Triantafillou, 1998). In this work, wood beams under three-point bending were reinforced by embedded bars of carbon-epoxy (Figure 6.18) with different heights to assess the gain obtained in stiffness and strength (Reis, 2013).

TABLE 6.3

Cohesive Parameters of Carbon-Epoxy and Structural Adhesive

Carbon-Epoxy				Structural Adhesive (SIKADUR 30)			
G_{Ic} (N/mm)	G_{IIc} (N/mm)	$\sigma_{1,I}$ (MPa)	$\sigma_{1,II}$ (MPa)	G_{Ic} (N/mm)	G_{IIc} (N/mm)	$\sigma_{1,I}$ (MPa)	$\sigma_{1,II}$ (MPa)
0.31	0.63	40	40	0.35	1.10	30	18

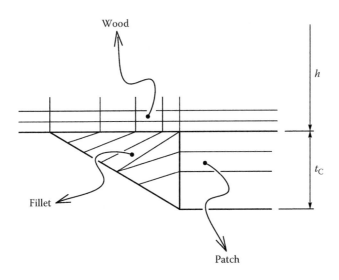

FIGURE 6.15
Mesh detail showing the adhesive fillet and several possible failure paths in wood, fillet and patch, identified by the straight lines (Dourado et al., 2012).

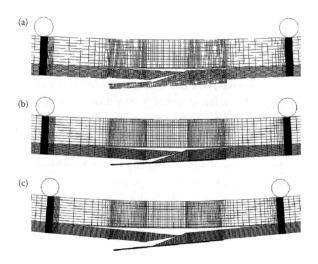

FIGURE 6.16
Crack paths obtained numerically considering (a) thick patch, (b) thin patch and (c) thin patch with fillet (Dourado et al., 2012).

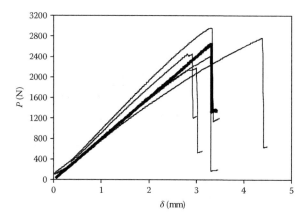

FIGURE 6.17
Numerical (thick line) and experimental $P-\delta$ curves for repaired beam with thin patch with fillet (Dourado et al., 2012).

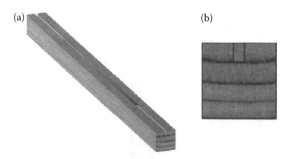

FIGURE 6.18
Schematic representation of the reinforced wood beam; (a) perspective and (b) front views.

Structural Applications – Case Studies 115

To validate the cohesive zone model, experimental tests were performed considering wood beams with a section of 20 × 20 mm² and a span of 300 mm (Figure 6.19). These beams were reinforced with carbon-epoxy bars with 200 mm length (L_C in Figure 6.19), width of 3 mm and a height of 5 mm.

The observed failure mode consisted of debonding of the carbon-epoxy bar followed by transverse fracture of wood beam (Figure 6.20). In addition, a remarkable indentation was also observed near the central loading point. This indentation reflects on non-linear behaviour of the load–displacement curve and should be accounted for in the numerical model. With this aim, wood plastic behaviour was considered using the stress–strain curves obtained by Reiterer et al. (2001) performing compression tests.

A three-dimensional finite element model was created considering cohesive zone elements placed in several positions: in the middle vertical section and between the carbon-epoxy bar and wood mimicking debonding. The simplest linear mixed-mode I + II + III cohesive law (Durão et al., 2006) was used. Fracture properties of wood and adhesive (SIKADUR® 30) are listed in Tables 6.1 and 6.3, respectively. Figure 6.21 reveals that the experimental failure mode (Figure 6.20) was accurately captured.

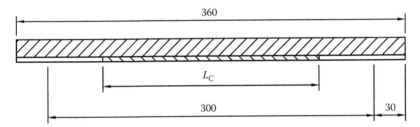

FIGURE 6.19
Dimensions of the reinforced wood beam.

FIGURE 6.20
Typical failure mode. (Adapted from Reis et al., 2018.)

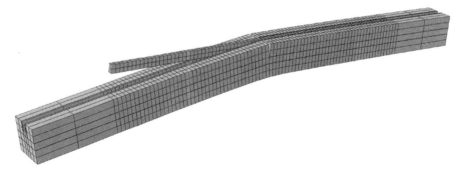

FIGURE 6.21
Failure modes observed in the finite element model.

In Figure 6.22, the numerical load–displacement curve is compared with the experimental ones. Globally, excellent agreement is obtained revealing the appropriateness of the model for this structural application.

The model was subsequently used to assess the influence of the reinforcement bar height on the stiffness (Figure 6.23) and load strength (Figure 6.24) of the wood reinforced beam. It should be noted that the value for an unreinforced beam is also included (reinforcement height equal to zero).

It can be observed that the stiffness and strength increase relatively to the unreinforced beam. A remarkable increase is obtained for reinforcement height of 3 mm. However, for higher values, only a marginal increase is achieved, meaning that the additional consumption of carbon-epoxy material is not compensated by structural improvement of the reinforced beam.

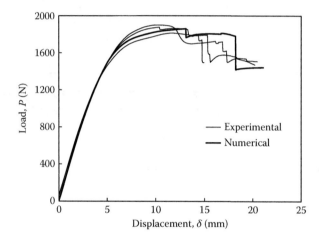

FIGURE 6.22
Numerical and experimental load–displacement curves. (Adapted from Reis et al., 2018.)

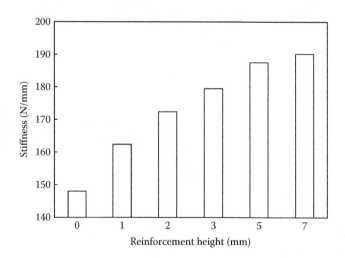

FIGURE 6.23
Evolution of specimen stiffness as a function of the reinforcement height.

Structural Applications – Case Studies

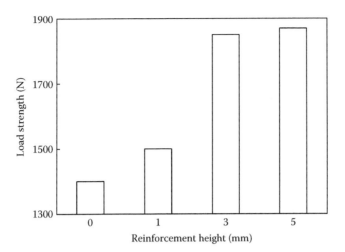

FIGURE 6.24
Evolution of specimen strength as a function of the reinforcement height.

This application is illustrative of the usefulness of the finite element analysis with cohesive zone modelling concerning the optimisation of structural details involving wood components. These models are able to account for damage development and to simulate failure modes observed experimentally. After initial validation, they can be used in systematic analysis of the influence of several parameters for the sake of structural optimisation, thus avoiding cumbersome and expensive experimental campaigns.

6.2 Wood Dowel Joints

Dowel joints are used due to their relevance on the improvement of strength and ductility. Load-carrying capacity of these joints is influenced by the joint geometry, material type and loading conditions. Load bearing capacity of timber structures is affected by the joints performance, being frequently the most critical point in a structure. In fact, these type of joints create geometrical discontinuities constituting sources of stress singularities in a structure, which are responsible for the global strength reduction.

6.2.1 Steel–Wood–Steel Connection

Cohesive zone modelling was applied to analyse the mechanical behaviour of a structural detail involving steel–wood–steel moment-resisting doweled joints submitted to bending. Figure 6.25 shows an arrangement formed by a central wood (*Pinus pinaster* Ait.) member and two external parallel metal (steel) plates connected by a pair of steel dowels (diameter D). The assembly is simply supported on two metal parts (cylinders) and submitted to a central load (P) at the top of both steel plates by means of a cylindrical device (Figure 6.26). Different arrangements were tested to analyse the influence of dowel spacing (L_1 in Figure 6.25) and dowel-to-end distance (L_2) on load-carrying capacity and initial stiffness (Caldeira et al., 2014). The loading line was kept at the mid-distance between both dowels regardless of the combination of the referred distances. Those measures were chosen as multiple values of the dowel diameter (D in Table 6.4) as follows: 3D, 5D and 7D,

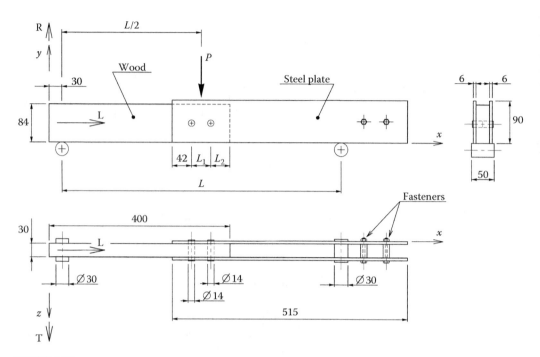

FIGURE 6.25
Dowel-joints showing wood anatomic directions: (L) Longitudinal, (R) Radial and (T) Tangential (Caldeira et al., 2014).

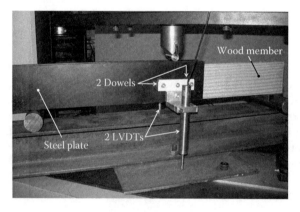

FIGURE 6.26
Three-point-bending test.

with $D = 14$ mm. In the experiments, two linear variable displacement transducers were used to capture the average displacement that was applied to the set of dowels relatively to the machine base. The mechanical tests were performed under displacement control at low rate (0.3 mm/min).

The numerical modelling of the referred testing arrangements was performed using the simplest linear cohesive mixed-mode I + II + III (Durão et al., 2006) damage model (Figure 4.7). Due to geometrical symmetry, only one-half of the specimen was modelled using adequate boundary conditions. Figure 6.27 shows a detail of the connection

Structural Applications – Case Studies

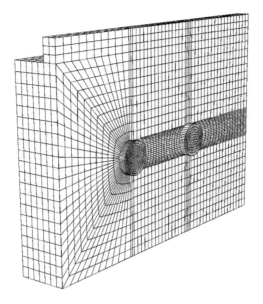

FIGURE 6.27
Detail of the FE mesh used in the connection region.

region where a refined mesh was used. Interface finite elements were judiciously positioned near the wood holes in several layers (diagonal crosses in Figure 6.28), which allows capturing the material failure with accuracy. Successive interface finite elements also allow constructing material damage mapping regions ensued by each dowel arrangement (Table 6.4). Analytical surfaces were considered to simulate the interface of the wood dowels, steel plate dowels and steel plate wood, thus preventing the material interpenetration during loading. Wood behaviour was simulated with the elastic properties of Tables 2.2 and 2.3, as well as the cohesive parameters summarised in Table 6.5. Dowels and metal plates were simulated as an isotropic material (E = 210 GPa and ν = 0.35).

Figure 6.29 shows details of the stress field (perpendicular to the grain) in the wood member (series 3D–3D) near the holes, with the developed cracks. These cracks have developed similar to the ones observed in the experiments (Figure 6.30). Figure 6.31a–i

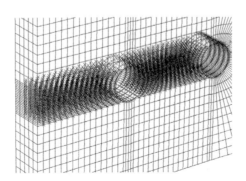

FIGURE 6.28
Detail of the FE mesh showing the position of interface finite elements near the holes (diagonal crosses) in different parallel layers.

TABLE 6.4

Dimensions (in mm) of Tested Arrangements according to Figure 6.25

Arrangements	L_1	L_2	L/2
3D–3D	42	42	307
3D–5D	42	70	279
3D–7D	42	98	251
5D–3D	70	42	293
5D–5D	70	70	265
5D–7D	70	98	237
7D–3D	98	42	279
7D–5D	98	70	251
7D–7D	98	98	223

TABLE 6.5

Cohesive Parameters of *Pinus pinaster* Ait. in the RL Fracture System (de Moura et al., 2009; Dourado et al., 2010, 2012)

G_{Ic} (N/mm)	G_{IIc} (N/mm)	$\sigma_{u,I}$ (MPa)	$\sigma_{u,II}$ (MPa)
0.26	0.91	5.34	9.27

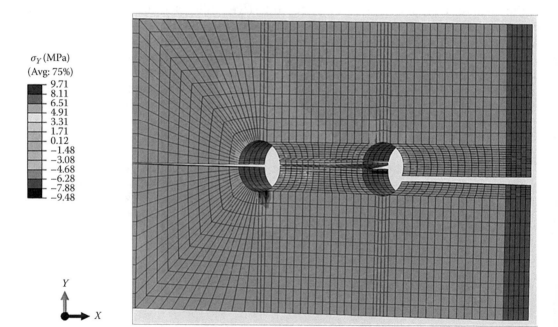

FIGURE 6.29
Stress field perpendicular to grain and developed cracks in the wood member (series 3D–3D).

illustrate the experimental load–displacement (P–δ) curves and the corresponding numerical computations obtained for series #D–3D, #D–5D and #D–7D (the symbol # is used to designate values 3, 5 and 7). Each curve was identified by the label L_1 and L_2, according to the notation presented in Table 6.4. Six or seven valid results per series

Structural Applications – Case Studies

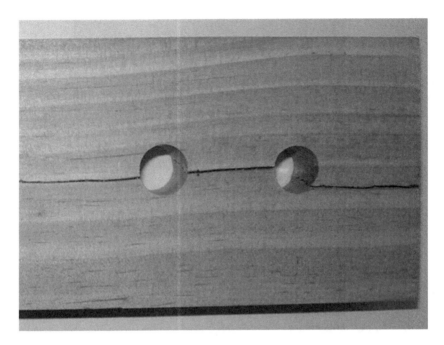

FIGURE 6.30
Typical damage observed experimentally in series 3D–3D.

were achieved experimentally. Besides replicating the experimental results coherently, both for the initial stiffness (Figure 6.32) and the ultimate load (Figure 6.33), the performed cohesive zone computations allow emphasising the effect of dowel-to-end distance (i.e. second index of the tested series in Table 6.4) on the mode failure. In effect, for series 3D–#D (i.e. Figure 6.31a–c), the pure brittle mode is attenuated as the dowel-to-end distance (second index of the series) increases from 3D to 7D, which was viewed as a beneficial measure to preclude the sudden catastrophic collapse of the connection. This behaviour was not observed in the remaining series (i.e. 5D–#D and 7D–#D in Figure 6.31d–i). Regarding the dowel spacing the numerical computations turned clear that a positive effect on the brittle failure modes is visible when a reduction of L_1 is made (see Figure 6.31i, f and c). These results, undoubtedly demonstrate that higher dowel-to-end distances and lower dowel spacing might be used to reduce brittle failure modes in this structural application. This conclusion is not in accordance with the statements of Eurocode 5 (European Committee for Standardization, 2004) that establishes the dimension of 7D as the minimum value to be used as the space between dowels (L_1) and to the boundary edge (L_2).

Figures 6.34a–c show the superposition of damage cartographies (fracture process zones and crack extent) for series 3D–#D, 5D–#D and 7D–#D, respectively, obtained for the displacement increment corresponding to the ultimate load. It becomes clear that a non-significant difference is visible when the dowel-to-end distance (parameter L_2 in Figure 6.25) is analysed for the same dowel-spacing measure (Figure 6.34a–c) for the referred displacement increment. Conversely, clear cracks (6.3 and 14 mm) are barely visible for series 5D–#D and 7D–#D, i.e. for wider distance dowel spacing.

6.2.2 Wood–Wood Joint

Standard solutions used to fabricate wooden members in timber construction often use adhesives to produce large cross sections with smaller but oriented boards. Before the arrival of structural adhesives, the wooden joints were fabricated by means of dowels and nails, frequently in low carbon steel alloys. In more distant times, those connecting elements were fabricated in tough wood species, tightly mounted. With the increasing interest

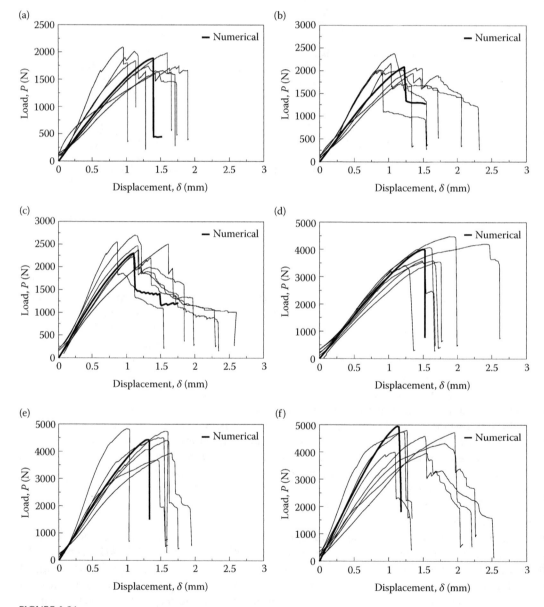

FIGURE 6.31
Numerical and experimental P–δ curves of tested series: (a) 3D–3D, (b) 3D–5D, (c) 3D–7D, (d) 5D–3D, (e) 5D–5D, (f) 5D–7D, (g) 7D–3D, (h) 7D–5D and (i) 7D–7D (Caldeira et al., 2014).

(Continued)

Structural Applications – Case Studies

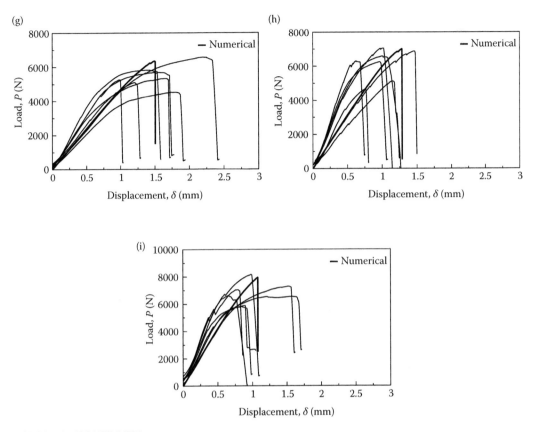

FIGURE 6.31 (CONTINUED)
Numerical and experimental P–δ curves of tested series: (a) 3D–3D, (b) 3D–5D, (c) 3D–7D, (d) 5D–3D, (e) 5D–5D, (f) 5D–7D, (g) 7D–3D, (h) 7D–5D and (i) 7D–7D (Caldeira et al., 2014).

for sustainable environmental product consumption, ancient fixing solutions in civil engineering may be recovered to reduce or even eliminate the use of structural adhesives that are harmful for the environment. Hence, a possible solution may involve the discrete disposition of wooden dowels throughout the contacting area, with an orientation that aims to assure the necessary stiffness and strength of the composed wood member. Figure 6.35a shows an illustration of a conceptual constructive model composed by two longitudinal (Figure 6.36) flat boards (*Pinus pinaster* Ait.) connected by means of four tilted wooden dowels (beech—*Fagus sylvatica*) fixed by an epoxy adhesive (SIKADUR 30 from Sika). The analysis involved wooden dowels symmetrically oriented relatively to the loading plane (Figure 6.35a) with four different inclinations θ (i.e. 30°, 45°, 60° and 90° in Table 6.6). A control model, used as a reference, was fabricated without dowels using a layer of adhesive to bond both boards (Figure 6.35b).

Quasi-static loading (Figure 6.37) was induced by means of three-point bending under displacement control, revealing the average load–displacement curves shown in Figure 6.38 for the referred series (five tests were performed for each case).

Figure 6.39 presents the attained results for the ultimate load (P_u) and the initial stiffness (K_0). These results allowed concluding that the beam stabilised with dowels tilted on 45° is the best solution for both the resistance and stiffness. Nevertheless, when compared with standard bonded solution, lower stiffness (−47%) and resistance (−20%) is obtained.

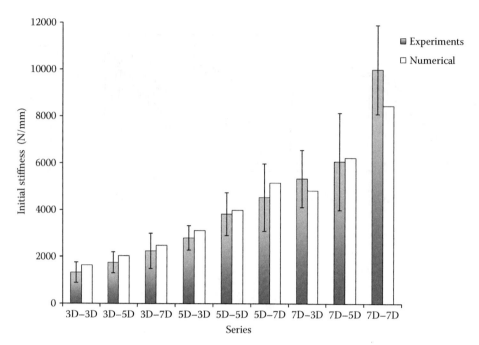

FIGURE 6.32
Mean values of initial stiffness and respective standard deviation.

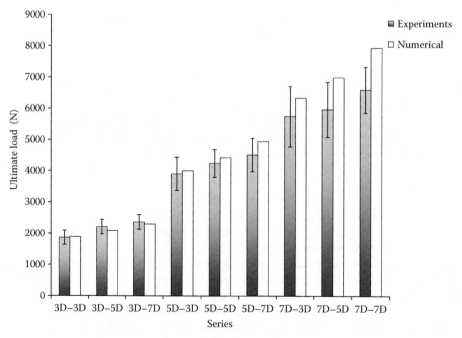

FIGURE 6.33
Mean values of ultimate load and respective standard deviation. (Adapted from Caldeira et al., 2014.)

Structural Applications – Case Studies

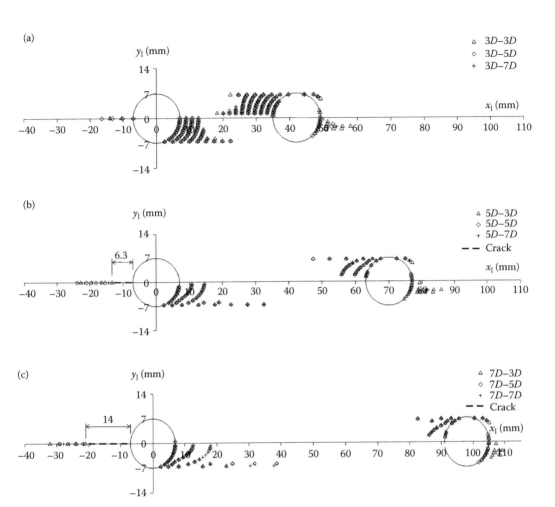

FIGURE 6.34
Damage extent (fracture process zones and crack extent) in the wood member close to the dowels, at ultimate load for series (a) 3D–#D, (b) 5D–#D and (c) 7D–#D.

The cohesive zone modelling of each assembly involved the construction of a finite element mesh with cohesive elements disposed along the loading plane (longitudinal-tangential (LT) fracture system of wood), where cracks arise (Figure 6.40). For the numerical model of the control series (i.e. bonded), the interface finite elements that allowed reproducing the mechanical behaviour of the adhesive layer were disposed along the area of contact of both boards (Figure 6.41a). For the remaining series, the interface finite elements were positioned along the border of the dowels (Figure 6.41b–e) that contact with the boards. The mechanical behaviour of the epoxy adhesive was replicated in a like manner as in Section 6.1.2 (Table 6.3). Pine wood (bulk) behaviour was simulated using the elastic properties shown in Tables 2.2 and 2.3, while the reproduction of the mixed-mode I + II failure (Figure 6.30) was made possible considering cohesive parameters that allowed replicating the non-linearity of the load–displacement curve. This procedure was justified by the fact that it is impossible to conceive a characterisation study that allows measuring the fracture toughness of wood perpendicular to the grain (namely,

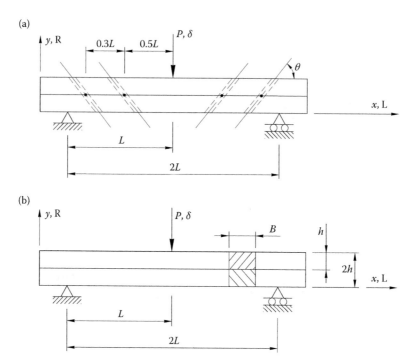

FIGURE 6.35
Wood composite beams: (a) stabilised by wood dowels and (b) bonded. $L = 190\,\text{mm}$, $h = 20\,\text{mm}$, $B = 75\,\text{mm}$.

TABLE 6.6

Specimens and Connecting Solution

Series	Connecting Solution	Specimens
B	Bonded	5
30	Dowels: 30°	5
45	Dowels: 45°	5
60	Dowels: 60°	5
90	Dowels: 90°	5

FIGURE 6.36
Specimen with 30° bent wood dowel.

Structural Applications – Case Studies

FIGURE 6.37
Experimental setup.

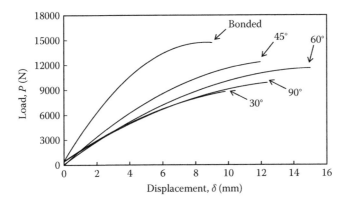

FIGURE 6.38
Average (five tests for each case) P–δ curves.

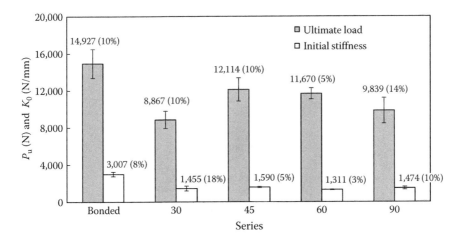

FIGURE 6.39
Average values for initial stiffness and ultimate load.

FIGURE 6.40
Crack propagation in a bonded specimen.

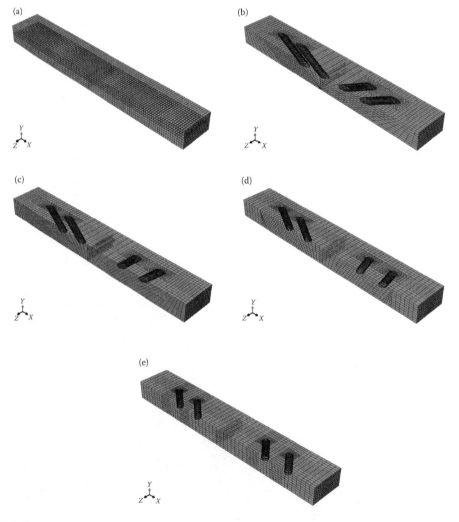

FIGURE 6.41
FE mesh showing the position of interface finite elements for series: (a) bonded, (b) tilted on 30°, (c) 45°, (d) 60° and (e) 90°.

Structural Applications – Case Studies

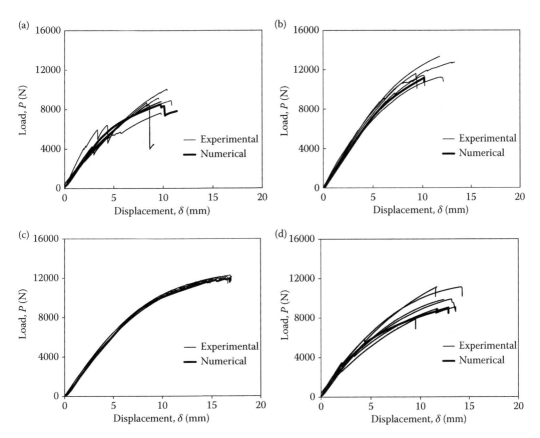

FIGURE 6.42
Numerical agreement of P–δ curves of series (a) 30°, (b) 45°, (c) 60° and (d) 90°.

the LT fracture system). Indeed, a fracture plane like the one shown in Figure 6.40 is only possible to occur in particular conditions, by means of a rigid device that forces wood to bend locally.

Figure 6.42a–d shows the general good agreement observed between numerical and experimental load–displacement curves obtained for series 30°, 45°, 60° and 90°, respectively.

Figure 6.43a,b illustrates the stress field at the peak-load near the loading device and crack propagation for series 30°. Damage obtained numerically is representative of the one observed experimentally. These figures show the appropriateness of the presented numerical methodology to replicate the damage profile in these wood structures.

6.3 Conclusions of Structural Applications

Five applications involving typical wood connections and structural details were used to illustrate the relevance of using non-linear fracture mechanics concepts via cohesive zone modelling in this context. Overall, good qualitative and quantitative representation of mechanical behaviour and failure modes were achieved, confirming that the proposed procedures can be viewed as fundamental tools aiming efficient design of wood structures.

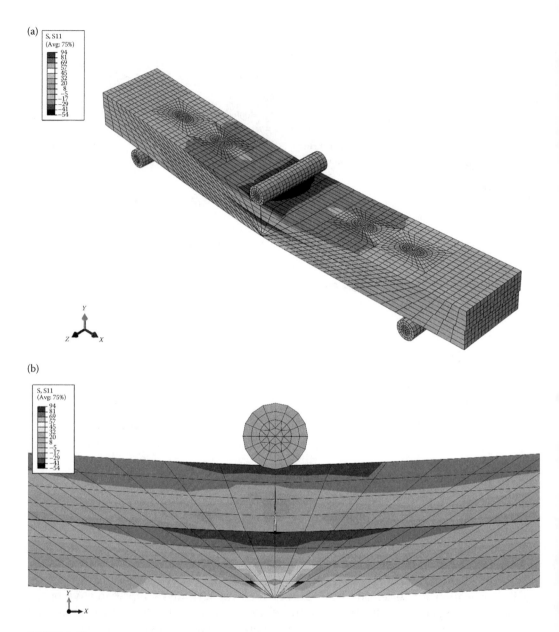

FIGURE 6.43
Stress distribution along the axial direction and crack propagation at the half-span for series 30° at the peak load.

References

Alhayek, H. and D. Svecova (2012). Flexural stiffness and strength of GFRP-reinforced timber beams. *J Compos Constr*, 16:245–52.

Barreto, A. M. J. P., R. D. S. G. Campilho, M. F. S. F. de Moura, J. J. L. Morais and C. L. Santos (2010). Repair of wood trusses loaded in tension with adhesively-bonded carbon-epoxy patches. *J Adhes*, 86:630–48.

Structural Applications – Case Studies

Borri, A., M. Corradi and A. Grazini (2005). A method for flexural reinforcement of old wood beams with CFRP materials. *Compos Part B*, 36:143–53.

Caldeira, T. V. P., N. Dourado, A. M. P. de Jesus, M. F. S. F. de Moura and J. J. L. Morais (2014). Quasi-static behavior of moment-carrying steel–wood doweled joints. *Constr Build Mater*, 53:439–47.

de Moura, M. F. S. F., M. A. L. Silva, J. J. L. Morais, A. B. de Morais and J. L. Lousada (2009). Data reduction scheme for measuring G_{IIc} of wood in end-notched flexure ENF tests. *Holzforschung*, 63:99–106.

Dourado, N., M. F. S. F. de Moura, J. J. L. Morais and M. A. L. Silva (2010). Estimate of resistance-curve in wood through the double cantilever beam test. *Holzforschung*, 64:119–26.

Dourado, N., F. A. M. Pereira, M. F. S. F. de Moura and J. J. L. Morais (2012). Repairing wood beams under bending using carbon-epoxy composites. *Eng Struct*, 34:342–50.

Durão, L. M., M. F. S. F. de Moura and A. T. Marques (2006). Numerical simulation of the drilling process on composites. *Compos Part A Appl Sci Manuf*, 37:1325–33.

European Committee for Standardization (2004). Eurocode 5: Design of timber structures—Part 1–1: General—Common rules and rules for buildings. EN 1995-1-1, Brussels.

Motlagh, Y., Y. Gholipour, G. Ebrahimi and M. Hosseinalibeygi (2008). Experimental and analytical investigations on flexural strengthening of old wood members in historical buildings with GFRP. *J Appl Sci*, 8:3887–94.

Reis, J. (2013). Estruturas de madeira reforçadas com compósitos, Tese Mestrado Integrado em Engenharia Mecânica, FEUP.

Reis, J. P., M. F. S. F. de Moura, F. G. A. Silva and N. Dourado (2018). Dimensional optimization of carbon-epoxy barsfor reinforcement of wood beams. *Compos Part B*, 139:163–70.

Reiterer, A. and S. E. S. Tschegg (2001). Compressive behaviour of softwood under uniaxial loading at different orientations to the grain. *Mech Mater*, 33:705–15.

Triantafillou, T. C. (1998). Composites: A new possibility for the shear strengthening of concrete, masonry and wood. *Compos Sci Technol*, 58:1258–95.

Index

Numbers

4ENF (four end-notched flexure test), 78–81
    CBBM (compliance-based beam method), 82–84
    compliance calibration method, 81–82
    experimental results, 84–87

A

adenosine triphosphate (ATP), 7
Anderson, T.L., 29
ATP (adenosine triphosphate), 7

B

Barrett, J.D., 6–7
Baurhoo, B., 7
bending loading, repaired beam under, 109–113
boats, 2
bonded joints
    repaired beam under bending loading, 109–113
    repaired beam under tensile loading, 105–109
bridges, 2

C

Caldeira, T.V.P., 117
Calvin cycle, 7
Carlsson, L.A., 74
case studies
    wood bonded joints
        reinforcement of wood structures, 113–117
        repaired beam under bending loading, 109–113
        repaired beam under tensile loading, 105–109
    wood dowel joints
        steel-wood-steel connection, 117–121
        wood-wood joint, 122–129
CBBM (compliance-based beam method)
    4ENF (four end-notched flexure test), 82–84
    DCB (double cantilever beam) test, 40–43
    ELS (end-loaded split test), 77–78
    ELS-MM (end load shear-mixed mode) test, 94–95

ENF (end-notched flexure test), 70–72
    numerical validation, 53–55
    SEN-TPB (single-edge-notched beam loaded in three-point-bending), 51–55
    single-leg bending (SLB) test, 91
    TDCB (tapered double cantilever beam), 58–60
cell differentiation, 9
cell lumen, 9
chairs, 2
chestnut, 5
classical data reduction schemes
    DCB (double cantilever beam) test, 37–39
    ELS (end-loaded split test), 76–77
    ENF (end-notched flexure test), 69–70
cohesive zone models (CZMs), 32–35, 42
combustion, 7
compact tension (CT) test, 37, 62–64
compliance calibration method, 4ENF (four end-notched flexure test), 81–82
compliance-based beam method (CBBM), *see* CBBM (compliance-based beam method)
constituents, wood, 7–10
Coté, A., 27
crack paths, 112–114
crack propagation, 128
Crews, J.H., 96
cross-laminated lumber, 3, 4
CT (compact tension) test, 37, 62–64
cypress, 5
CZMs (cohesive zone models), 32–35, 42, 73

D

da Silva, Manuel António Lima, ix
damage extent, 125
data reduction scheme based on equivalent LEFM, 49–51
data reduction scheme based on TDCB, 55–58
DCB (double cantilever beam) test, 37–40
    CBBM (compliance-based beam method), 40–42, 40–43
    experimental results, 45–47
    MECM (modified experimental compliance method), 38–40
    numerical results, 45–47
    numerical validation, 42–45

134 *Index*

de Moura, Marcelo, xi, 68
de Oliveira, Jorge Marcelo Quintas, ix
diffuse-porous woods, 11
Dohertya, W.O.S., 7
double cantilever beam (DCB) test, *see* DCB
(double cantilever beam) test
Douglas fir, 5
Dourado, Nuno, xi, 109
Durão, L.M., 115, 118

E

earlywood, 10–11
eastern hemlock, 5
elastic properties, 15–26
pine wood and carbon-epoxy laminate, 108
elm, 5
ELS (end-loaded split test), 74–76
CBBM (compliance-based beam method),
77–78
classical data reduction schemes, 76–77
experimental results, 78–79
ELS-MM (end load shear-mixed mode) test,
93–94
CBBM (compliance-based beam method),
94–95
experimental results, 96–97
numerical analysis, 95–96
EMC (equilibrium moisture content), 12–13
end load shear-mixed mode (ELS-MM) test, *see*
ELS-MM (end load shear-mixed mode)
test
end-notched flexure test (ENF), *see* ENF (end-
notched flexure test)
ENF (end-notched flexure test), 68–69
CBBM (compliance-based beam method),
70–72
classical data reduction schemes, 69–70
experimental results, 72–74
equilibrium moisture content (EMC), 12–13
equivalent LEFM, SEN-TPB (single-edge-
notched beam loaded in three-point-
bending), 49–51
experimental results
4ENF (four end-notched flexure test), 84–87
DCB (double cantilever beam) test, 45–47
ELS (end-loaded split test), 78–79
ELS-MM (end load shear-mixed mode) test,
96–97
ENF (end-notched flexure test), 72–74
MMB (mixed-mode bending) test, 99–102
single-leg bending (SLB) test, 93
TDCB (tapered double cantilever beam), 61–62

F

failure modes, 115
failure paths, 110–111
FEA (finite element analysis), mechanical
properties, 43–44
finger-jointed lumber, 3, 4
finite element analysis (FEA), 43–44
four end-notched flexure test (4ENF), *see* 4ENF
(four end-notched flexure test)
FPM (fracture process zone), 39
fracture process zone (FPZ), 32, 39
fracture systems, 38

G

glued laminated timber, 3, 4
growing points, 5
growth rings, 11
Gustafsson, P.J., 47

H

Hankinson formulas, 27
Harlow, W.M., 10
Harrar, E.S., 10
harvesting of wood, 3
heartwood, 5, 8
hemicelluoses, 7
Herrmann, G., 51
hockey stick, 2
Hooke's law, 16

I

I-beams, 3, 4
I-joists, 3, 4
Iosipescu test, 25
Irwin, G.R., 30
Irwin-Kies equation, 69–70
Irwin-Kies expression, 37, 60
Irwin-Kies relation, 30

K

Kienzler, R., 51
Kies, J.A., 30
linear elastic fracture mechanics (LEFM),
29–32
Krueger, R., 31

L

laminated strand lumbar, 3, 4
laminated veneer lumber, 3, 4

Index

latewood, 10–11
leaves, 5–7
LEFM (linear elastic fracture mechanics), 29–32
  equivalent LEFM, data reduction scheme based on, 49–51
Leonardo bridge, 2
linear elastic fracture mechanics (LEFM), 29–32
linear regression, applied torsional moment, 22
Liu, J.Y., 28
Lousada, José, ix

M

MC (moisture content), 12–13
MECM (modified experimental compliance method), DCB (double cantilever beam) test, 39–40
medium-density fibreboard, 3–4
micro-structure, wood, 7–10
mixed-mode bending (MMB) test, *see* MMB (mixed-mode bending) test
mixed-mode I + II fracture characterization, 89, 103
  ELS-MM (end load shear-mixed mode) test, 93–94
    CBBM (compliance-based beam method), 94–95
    experimental results, 96–97
    numerical analysis, 95–96
  MMB (mixed-mode bending) test, 97–99
    experimental results, 99–102
    numerical validation, 102
  single-leg bending (SLB) test, 90
    CBBM (compliance-based beam method), 91
    experimental results, 93
    numerical analysis, 91–93
mixed-mode I + II trapezoidal with bilinear softening cohesive law, 32
MMB (mixed-mode bending) test
  experimental results, 97–102
  numerical validation, 102
mode I fracture characterization, 37
  compact tension (CT) test, 62–64
  conclusions, 64
  DCB (double cantilever beam) test, 37–40
    CBBM (compliance-based beam method), 40–42, 40–43
    experimental results, 45–47
    MECM (modified experimental compliance method), 38–40

numerical results, 45–47
numerical validation, 42–45
SEN-TPB (single-edge-notched beam loaded in three-point-bending), 47–49
  CBBM (compliance-based beam method), 51–55
  equivalent LEFM, 49–51
TDCB (tapered double cantilever beam), 55
  CBBM (compliance-based beam method), 58–60
  data reduction scheme based on, 55–58
  experimental results, 61–62
  numerical validation, 60–61
mode II fracture characterization, 67, 87
  4ENF (four end-notched flexure test), 79–81
    CBBM (compliance-based beam method), 82–84
    compliance calibration method, 81–82
    experimental results, 84–87
  ELS (end-loaded split test), 74–76
    CBBM (compliance-based beam method), 77–78
    classical data reduction schemes, 76–77
    experimental results, 78–79
  ENF (end-notched flexure test), 68–69
    CBBM (compliance-based beam method), 70–72
    classical data reduction schemes, 69–70
    experimental results, 72–74
modified experimental compliance method (MECM), *see* MECM (modified experimental compliance method)
moisture, 12–13
moisture content (MC), 12–13
Morais, José Joaquim Lopes, ix
Morel, Stéphane, ix
Mostovoy, S., 55, 60

N

Norris, C.B., 28
numerical analysis
  ELS-MM (end load shear-mixed mode) test, 95–96
  single-leg bending (SLB) test, 91–93
numerical results
  4ENF (four end-notched flexure test), 84–87
  DCB (double cantilever beam) test, 45–47
  ELS (end-loaded split test), 78–79
  ENF (end-notched flexure test), 72–74
  TDCB (tapered double cantilever beam), 61–62

136

numerical validation
  CBBM (compliance-based beam method),
    53–55
  DCB (double cantilever beam) test, 42–45
  MMB (mixed-mode bending) test, 102
  TDCB (tapered double cantilever beam),
    60–61

O

oak, 5
OD (oven drying), 12
off-axis tensile test, 24
Olivera, J.M., 94–95
oriented strand board, 3–4
orthotropic directions, 15
oven drying (OD), 12

P

paper, 3–4
parallel strand lumber, 3–4
particleboard, 3–4
photosynthesis, 7
pine, 5
pine wood and carbon-epoxy laminate,
    elastic properties, 108
plant cell, 8–9
plasma membrane, 9
plywood, 3–4
Poisson's ratios, 16, 20
port wine carrying boat, 2
Puukoukka building, 3

Q

quasi-static loading, 123

R

redwood, 5
Reeder, J.R., 96
reinforcement of wood structures, 113–117
Reis, João Pedro da Costa, ix
Reiterer, A., 115
Rivers, S., 12
roots, 5–7

S

sapwood, 5
SEN-TPB (single-edge-notched beam loaded in
    three-point-bending) test, 37, 47–49

CBBM (compliance-based beam method),
    51–55
equivalent LEFM, data reduction scheme
    based on, 49–51
shape coefficients, shear test, 22
shear moduli, 20–26
single-edge-notched beam loaded in three-
    point-bending (SEN-TPB) test, 37
single-leg bending (SLB) test, 90
  CBBM (compliance-based beam method), 91
  experimental results, 93
  numerical analysis, 91–93
SLB (single-leg bending) test, *see* single-leg
    bending (SLB) test
softwoods, formation, 10
steel-wood-steel connection, wood dowel joints,
    117–121
stems, 5–7
strength failure criteria, 26–29
strength properties, 15–26
  Poisson's ratios, 20
  shear moduli, 20–26
  tensile tests, 17
  Young's moduli, 16–20
stress distribution, 130

T

tapered double cantilever beam (TDCB) test, 37
TDCB (tapered double cantilever beam) test,
    37, 55
  CBBM (compliance-based beam method),
    58–60
  data reduction scheme based on, 55–58
  experimental results, 61–62
  numerical validation, 60–61
tensile joints, repaired beam under, 105–109
tensile tests, specimen geometries and
    dimensions, 17
tests
  4ENF (four end-notched flexure test), 78–81
    CBBM (compliance-based beam method),
      82–84
    compliance calibration method, 81–82
    experimental results, 84–87
  compact tension (CT) test, 62–64
  DCB (double cantilever beam) test, 37–40
    experimental results, 45–47
  ELS (end-loaded split test), 74–76
    CBBM (compliance-based beam method),
      77–78
    classical data reduction schemes, 76–77
    experimental results, 78–79

Index

ELS-MM (end load shear-mixed mode) test, 93–94
  CBBM (compliance-based beam method), 94–95
  experimental results, 96–97
  numerical analysis, 95–96
ENF (end-notched flexure test), 68–69
  CBBM (compliance-based beam method), 70–72
  classical data reduction schemes, 69–70
  experimental results, 72–74
MMB (mixed-mode bending) test, 97–99
  experimental results, 99–102
  numerical validation, 102
SEN-TPB (single-edge-notched beam loaded in three-point-bending), 47–49
  CBBM (compliance-based beam method), 51–55
  equivalent LEFM, 49–51
shear test, 22
single-leg bending (SLB) test, 90
  CBBM (compliance-based beam method), 91
  experimental results, 93
  numerical analysis, 91–93
TDCB (tapered double cantilever beam), 55
  CBBM (compliance-based beam method), 58–60
  data reduction scheme based on, 55–58
  experimental results, 61–62
  numerical validation, 60–61
timber frames, 2
Timoshenko beam theory, 40
torsional test, 21
Tsai, S.W., 28
Tsai-Hill criterion, 27
tyloses, 5

U

ultimate load, 124
Umney, N., 12

V

Valentin, Gérard, ix

W

walnut, 5
wood
  constituents, 7–10
  energy source, 3
  formation, 5–7
  harvesting, 3
  mesoscale, 10–13
  micro-structure, 7–10
  moisture, 12–13
  products, 3–4
  softwoods, 10
  uses, 1
wood bonded joints
  reinforcement of wood structures, 113–117
  repaired beam under bending loading, 109–113
  repaired beam under tensile loading, 105–109
wood dowel joints
  steel-wood-steel connection, 117–121
  wood-wood joint, 122–129
wood I-beams, 3, 4
wood mechanical behavior
  cohesive zone models (CZMs), 32–35
  elastic properties, 15–26
  strength failure criteria, 26–29
  strength properties, 15–26
    Poisson's ratios, 20
    shear moduli, 20–26
    tensile tests, 17
    Young's moduli, 16–20
wooden furniture, 2
wood-wood joint, wood dowel joints, 122–129
Wu, E.M., 28

X

xylem, 5

Y

Yoshihara, H., 20–26, 67–68, 79
Young's moduli, 16–17, 81

PGSTL 06/13/2018